THE NUTRIENT ROADMAP

WEF Special Publication

2015

Water Environment Federation
601 Wythe Street
Alexandria, VA 22314–1994 USA
http://www.wef.org

ISBN 978-1-57278-314-0

About WEF

Founded in 1928, the Water Environment Federation (WEF) is a not-for-profit technical and educational organization of 36,000 individual members and 75 affiliated Member Associations representing water quality professionals around the world. WEF members, Member Associations and staff proudly work to achieve our mission to provide bold leadership, champion innovation, connect water professionals, and leverage knowledge to support clean and safe water worldwide. To learn more, visit www.wef.org.

Prepared by the Nutrient Roadmap Task Force of the Water Environment Federation

Samuel S. Jeyanayagam, Ph.D., P.E., BCEE, *Chair*

Sara Arabi, Ph.D., P.Eng.
Madan Arora, Ph.D., P.E., BCEE
Chuck Berger
Lisa Boudeman
John Bratby, Ph.D., P.E.
Jeanette Brown, P.E., BCEE, D. WRE, F.WEF
David G. Bunch, P.E.
Robin Burch
Dennis Burke
Jennifer Callahan
Chein-Chi Chang, Ph.D., P.E.
S. Rao Chitikela, Ph.D., P.E., P.Eng., BCEE
Erik R. Coats, P.E., Ph.D.
Francisco J. Cubas, Ph.D.
Krzysztof Czerwionka
Rony Das
Christine DeBarbadillo
Rowland C. Denny
John Donovan, P.E., BCEE
Douglas Drury, Ph.D.
Nico Enthoven
Zeynep Erdal, Ph.D., P.E.
Eric Evans
Michael Falk, Ph.D., P.E.
Richard Finger
James D. Fitzpatrick, P.E.
Dawn Flancher, P.E.
Dave Flowers
Val S. Frenkel, Ph.D., PE, D.WRE
Elisa Garvey, Ph.D., P.E.
Joseph M. Gianvito, P.E.
Matthew Gregg, P.E.
Marek Gromiec
Scott Gulliams
Mohamed Hamoda, Ph.D., P.Eng.
Brent Herring
Hans Holmberg
Dwight Houweling, Ph.D., P.Eng.
Joseph A. Husband, P.E., BCEE
Sidney Innerebner, Ph.D., P.E., CWP
Seshadri S. Iyer, Ph.D., P.E., CFM, CNMP
Gary R. Johnson, P.E., BCEE
Ishin Kaya, P.Eng
Brandon Koltz
Hua (Larry) Li, Ph.D., P.E.
Craig Lindell
Jorj Long
Jingwei Ma, Ph.D.
Jose Christiano Machado Jr., Ph.D., P.E.
Maritza Macias-Corral, Ph.D.
Jane Madden, P.E., BCEE
Scott McClelland
Indra N. Mitra, Ph.D., P.E., MBA, BCEE
Timothy Mussen, Ph.D.
Weihua (Peter) Peng, Ph.D., P.E., BCEE
Heather M. Phillips, P.E., BCEE
Anne Rea
Rod Reardon
Joe Rudek
Kevin Samson
Randal W. Samstag, M.S., P.E., BCEE
Devendra Saroj, Ph.D.

Robert R. Sharp, Ph.D., P.E., BCEE
Michael Simpson
Kyong Song
Eric Spargimino, P.E., LEED AP
Paul E. Stacey
Carl Stone, P.E., BCEE
Michael B. Tate, P.E.
Lisa Thompson, Ph.D.
Ya-Chi Tsao, Ph.D., P.E.
Amanda Vincent
Marc Walch
Justin J. Waples, P.E., ENV SP
Brett Ward
Grant Weaver, P.E.
Jianfeng Wen
Tami Wetmore
David Winters
Wayne Wong, M.A.Sc., P.Eng., PMP
Qingzhong Wu, Ph.D., P.E.
Yu Yang
Harry Zhang, Ph.D., P.E.

Under the Direction of the Municipal Design Subcommittee of the Technical Practice Committee

2015

Water Environment Federation
601 Wythe Street
Alexandria, VA 22314-1994 USA
http://www.wef.org

Special Publications of the Water Environment Federation

The WEF Technical Practice Committee (formerly the Committee on Sewage and Industrial Wastes Practice of the Federation of Sewage and Industrial Wastes Associations) was created by the Federation Board of Control on October 11, 1941. The primary function of the Committee is to originate and produce, through appropriate subcommittees, special publications dealing with technical aspects of the broad interests of the Federation. These publications are intended to provide background information through a review of technical practices and detailed procedures that research and experience have shown to be functional and practical.

Contents

Chapter 4 Resource Recovery
Jane Madden, P.E., BCEE

List of Figures

List of Tables

Preface

Nutrients (nitrogen and phosphorus) are found in agricultural and home fertilizers as well as in livestock, industrial, and municipal wastes. Population growth, urbanization, and improved quality of life have increased the demand for fertilizers and the combustion of fossil fuels. Though nutrients are essential to life, when present in excess, they are harmful water pollutants. It has been estimated that because of inefficiencies in the production, handling, and use, a significant amount of newly produced nitrogen and phosphorus is lost to the environment. The resulting environmental damage, quantified by the U.S. Environmental Protection Agency, represents more than 14,000 nutrient impaired water bodies and more than 168 hypoxic (low dissolved oxygen) zones in the United States.

To address this issue, in October 2013, the Water Environment Federation (WEF), Environmental Defense Fund (EDF), and Johnson Foundation at Wingspread sponsored a meeting in Airlie, Virginia (the "Airlie meeting"), to craft a nutrient management vision and initiate the development of a roadmap for realizing that vision. The Water Environment Federation's mission is to "provide bold leadership, champion innovation, connect water professionals, and leverage knowledge to support clean and safe water worldwide". The Environmental Defense Fund's mission is to "preserve the natural systems on which all life depends and, guided by science and economics, find lasting solutions to the most serious environmental problems". The Johnson Foundation at Wingspread's mission is to "be a catalyst for positive and lasting change leading to healthier environments and communities". The Airlie meeting brought together 28 diverse representatives from water utilities, regulatory agencies, academia, consultants, associations, and nongovernmental organizations to develop the following draft aspirational goal:

> The next generation of wastewater treatment has zero net impact with regards to energy use, greenhouse gas emissions and nutrient discharge by 2040. Achieving this goal will require a dedication to overcoming the technical barriers, financial constraints and regulatory disincentives limiting nutrient removal, [greenhouse gas] emission reduction and energy neutrality in the treatment of wastewater.

Clearly, the group's consensus is that to remain sustainable, disruptive changes are needed in how we manage our precious and finite resources.

The Nutrient Roadmap, prepared by industry experts, is a significant step towards accelerating the transition to smarter nutrient management.

The Nutrient Roadmap is a high-level overview document that provides guidance to utility managers, decision-makers, and other interested parties. (For an in-depth technical discourse, the reader is directed to other publications, including WEF Manuals of Practice and Special Publications.) With the help of *The Nutrient Roadmap*, the reader should be able to identify where the utility is with respect to controlling nutrient discharges and the next steps in achieving sustainable nutrient management in the context of the utility's specific regulatory climate. The manual also attempts to identify future research, training, and advocacy needs to support the move towards smarter nutrient management and enable the transformation to the water resource recovery facility of the future. This publication was produced under the direction of Samuel S. Jeyanayagam, Ph.D., P.E., BCEE, *Chair*.

Authors' and reviewers' efforts were supported by the following organizations:

3M Company, Cottage Grove, Minnesota
AECOM, Burnaby, British Columbia, Canada
Atlantic South Consulting Services LLC, Charleston, South Carolina
American Water Works Association, Denver, Colorado
ARCADIS, Inc., White Plains, New York
Black & Veatch, Corporation, St. Louis, Missouri; Cincinnati, Ohio; and Kansas City, Missouri
Brentwood Industries, Reading, Pennsylvania
Brown and Caldwell, Walnut Creek, California
CDM Smith Inc., Boston, Massachusetts, and Denver, Colorado
CH2M HILL, Chantilly, Virginia; Santa Ana, California; and Toronto, Ontario, Canada
City of Olathe, Kansas
Clark County Water Reclamation District, Las Vegas, Nevada
CONACYT-Cátedras, CIMAV Unidad Durango, México
Connecticut Department of Energy and Environmental Protection, Hartford, Connecticut
Carollo Engineers, Walnut Creek, California
D.C. Water and Sewer Authority, Washington, D.C.
Environmental Defense Fund, New York, New York
Environmental Operating Solutions, Inc. (EOSi), Bourne, Massachusetts
Georgia Southern University, Statesboro, Georgia
HDR, Folsom, California
Indigo Water Group, LLC, Littleton, Colorado

Johnson Controls, Inc., Westerville Ohio
Kansas Department of Health and Environment, Topeka, Kansas
KLH Engineers, Inc., Pittsburgh, Pennsylvania
Kuwait University, Kuwait
LimnoTech, Ann Arbor, Michigan
New Hampshire Fish & Game Department, Great Bay National Estuarine Research Reserve, Greenland, New Hampshire
Philadelphia Water Department, Philadelphia, Pennsylvania
Randal W. Samstag Civil and Sanitary Engineer, Bainbridge Island, Washington
Sacramento Regional County Sanitation District, Sacramento, California
The Water Planet Company, New London, Connecticut
UEM Group (a Toshiba affiliated Company), India
United Water Services, Indianapolis, Indiana
University of Idaho, Moscow, Idaho
University of Surrey, Surrey, U.K.
Washington State University, Pullman, Washington
Water & Environmental Consulting LLC, Milwaukee, Wisconsin
Weston & Sampson, Peabody, Massachusetts

Executive Summary

The Nutrient Roadmap was written to guide decision-makers towards fully sustainable wastewater treatment by providing opportunities for phosphorus and nitrogen recovery and reuse, rather than simply removal. The publication does not set out a singular path; instead, it provides decision-makers with various routes to consider while offering alternative destinations from which to choose.

Those who use the roadmap are encouraged to retool their facilities into water resource recovery facilities (WRRFs) that not only produce clean water, but recover nutrients in an energy-neutral manner consistent with Water Environment Federation's (WEF's) 2014 publication, *Moving Toward Resource Recovery Facilities*.

A sustainable approach to managing nutrients involves tradeoffs. Decision-makers are encouraged to consider environmental and community effects, operational effectiveness, and economic factors. Water resource recovery facility managers are encouraged to maximize the capture and reuse of waste-stream nutrients, minimize the energy used to process wastewater, reduce the release of nutrients into the environment, minimize the release of greenhouse gas emissions, and maximize economic benefits.

Innovative, cost-effective solutions for permit compliance are identified. Proactive actions in anticipation of new or more stringent regulatory requirements are explained, including those that reduce operational costs. Utilities not currently subject to strict nutrient discharge limits are encouraged to prepare for the future by adopting some of the pioneering practices that highly regulated WRRFs are undertaking.

A number of treatment technologies are available for both mainstream and sidestream treatment of total nitrogen and total phosphorus. Details of these technologies can be found in WEF Manual of Practice (MOP) No. 8, *Design of Municipal Wastewater Treatment Plants*; WEF MOP No.

11, *Operation of Municipal Wastewater Treatment Plants*; and WEF MOP No. 34, *Nutrient Removal.*

As discussed in this publication, operational changes can oftentimes create the necessary biological habitats to allow for nutrient removal. When possible, such an approach provides for highly sustainable nutrient management.

Water resource recovery facilities can achieve low nutrient discharges through a variety of processes, including biological nutrient removal (BNR), physical separation, and chemical methods. In selecting an approach, economic and environmental trade-offs exist. Most technologies capable of removing both nitrogen and phosphorus use BNR, which relies on biota to transform the nutrients present in wastewater to gaseous or insoluble (particulate) forms. Achieving significant nutrient reduction requires a greater emphasis on data analysis and process control to optimize the environmental conditions that sustain nutrient-removing organisms.

In some cases, BNR is followed by chemical methods to further reduce total phosphorus concentrations. *Mainstream nutrient* treatment takes place within the typical process flow. *Sidestream treatment* refers to liquid resulting from biosolids processing that is intercepted with the goal of removing nutrients from a concentrated stream; like mainstream nutrient treatment processes, sidestream treatment also can vary from biological to physical and chemical removal methods.

Technologies currently exist to allow facilities to recover nutrients in pelletized form (struvite). Although it is typically not cost-effective for smaller treatment facilities to recover struvite for commercial use, even the smallest of WRRFs generates a product that is nutrient-rich and, therefore, of value.

Struvite pellets, which contain nitrogen and phosphorus, can, and are, being beneficially used as a fertilizer. The land application of biosolids, when practiced properly, also provides for beneficial reuse of nutrients.

MATRICES

The matrices found in the Appendix provide a high-level overview of the roadmap to net-zero nutrient discharges. Under each topic, objectives are outlined following three phases, from planning to implementation to evaluation and improvement.

1

Introduction

1.0 PURPOSE OF *THE NUTRIENT ROADMAP*

1.1 Timing of *The Nutrient Roadmap*

Historically, wastewater treatment operators have managed nutrients, nitrogen, and phosphorus removal from the water at any cost, trying to reach the lowest possible concentrations in the treated water to minimize the effect on the environment and to comply with water discharge regulations. The tighter the treatment, the higher the cost. While operators previously looked at nutrients as unwanted pollutants that need to be removed and disposed of, that view has changed.

1.2 From Wastewater Treatment to Resource Recovery

The wastewater treatment industry learned that it can intensify natural methane gas production from wastewater, gain energy, and remove nutrients by using new advanced and emerging technologies. However, it did not explore deeply opportunities to turn the nutrients removal process to the resource recovery concept. When this happens, benefits will be gained from the recovered nutrients, reducing needs and the cost of treatment and disposal. According to the Water Environment Federation (WEF) 2011 Renewable Energy Position Statement:

> Wastewater treatment plants are not the waste disposal facilities, but are the water resource recovery facilities that produce clean water, recover nutrients such as phosphorus and nitrogen, and have the potential to reduce the nation's dependence on fossil fuels through the production and use of renewable energy and the implementation of energy conservation.

2.0 DRIVERS FOR CHANGE

2.1 Population Growth

Population growth and corresponding increases in the demand for food and energy have driven the use of phosphorus and nitrogen-based fertilizers and the combustion of fossil fuels. These factors have contributed to the rapid dissemination of reactive nitrogen in atmospheric, terrestrial, and aquatic environments.

2.2 Environmental Concerns

Nitrogen overloading causes cultural eutrophication, which is the excessive richness of nutrients in a lake or other body of water, frequently because of the following:

- Runoff from the land, which causes a dense growth of plant life and death of animal life from lack of oxygen, and biodiversity loss from dead zones in estuaries and freshwater systems;
- Increased pollutants and potent greenhouse gas emissions in atmospheric systems; and
- Through atmospheric deposition, increased acidification and other forms of nitrogen pollution in soils.

Excessive concentration of ammonia and orthophosphates is toxic for the habitats of rivers and lakes. Similarly, too much phosphorus and nitrogen in waters contribute to toxic algal blooms and eutrophication in lakes and other aquatic environments, causing significant environmental damage. Rapid depletion of natural rock phosphate, the main source of phosphorus for industry for human needs, attracts nutrient recovery to allow preservation of natural resources.

2.3 Clean Water

At the same time that society is demanding cleaner water, new opportunities are being developed to remove nutrients from wastewater more effectively and to recover them. Although many nutrients can be costly or energy-intensive to remove, they can also be revenue-generating or at least cost-offsetting when recovered. Recovery of the significant quantities of nitrogen and phosphorus in municipal waste streams could help to stem the growing nutrient imbalances being observed in natural systems, offer a marketable product, and, in the case of phosphorus, provide a sustainable source of a finite resource to support agricultural production and food security in the United States and worldwide.

3.0 REGULATORY ASPECTS

In 2013, WEF, the Water Environment Research Foundation, and the National Association of Clean Water Agencies released *The Water Resources Utility of the Future: A Blueprint for Action* to define relevant issues, analyze key data, and offer recommendations for critical actions for the future of water resources management. Today's clean water agencies are increasingly looking beyond the Clean Water Act (CWA) as they seek to improve environmental performance, benefit their communities, and improve their financial picture. The "water resources utility of the future" is defined by today's utility leaders who are pioneering innovative technologies and cutting-edge practices with a focus on resource recovery. The aforementioned publication presents objectives and goals for the water community to adopt in an effort to achieve a new clean water paradigm that considers energy production, water and biosolids reuse, green infrastructure, and watershed-based approaches. Under section 303(d) of the CWA, U.S. states, territories, and authorized tribes are required to develop lists of impaired waters. The law requires that these jurisdictions establish priority rankings for waters on the lists and develop total maximum daily loads for these waters.

4.0 FUTURE TRENDS

Water Environment Federation and the Environmental Defense Fund (EDF) both recognize the value of incentivizing technology improvements for removal and/or recovery and reuse of nitrogen and phosphorus from wastewater in a sustainable and cost-effective way. Moreover, the recovered nutrients may be the revenue generators in the future, complementing other benefits listed previously. The two organizations jointly crafted the following vision statement for the future of wastewater treatment: "WEF and EDF are collaboratively seeking to explore 'smarter' approaches to the management of nutrients from municipal water sources, considering such factors as the potential for resource recovery, overall net environmental benefit, regulatory barriers/incentives and cost effectiveness".

Technology innovation not only has to be incentivized, but also the technologies for the decision-making processes that assess the risks before the decisions are made and audit the decisions for their effectiveness and consequences after they are implemented. The public decision-making process must be disciplined and adaptive to match the aspirations of the proposed program. It should include source reduction programs such as the reduction of fertilizer use and runoffs to the receiving waters. Finally, the wastewater treatment industry should start implementing the resource recovery programs with existing operational facilities and plan resource recovery programs for new facilities.

5.0 REFERENCES

National Association of Clean Water Agencies; Water Environment Research Foundation; Water Environment Federation (2013) *The Water Resources Utility of the Future: A Blueprint for Action;* Water Environment Research Foundation: Alexandria, Virginia.

Water Environment Federation; Environmental Defense Fund (2013) Joint Vision Statement. http://www.wef.org/nutrientroadmap/.

Water Environment Federation (2011) *Renewable Energy Position Statement*; Water Environment Federation: Alexandria, Virginia.

2

Setting the Stage for Nutrient Management

1.0 DEVELOPING THE VISION FOR NUTRIENT REMOVAL

For municipal and utility managers to develop a comprehensive plan for nutrient management, they must understand the drivers, regulatory requirements, and effects of nutrients both within the affected watershed and beyond.

1.1 Drivers

The primary drivers for controlling nutrient discharges include improving water quality, protecting public and environmental health, and enhancing or maintaining economic benefit. The discharge of excessive nutrient loads to waterbodies can result in a variety of biological responses that may cause algal blooms, choking weeds, dissolved oxygen depletion, taste and odor problems, turbidity, decreases in biodiversity, and elimination of important species. The result is an impaired waterbody and possible effects on economic activity such as commercial and recreational fishing and shellfish harvesting, recreational swimming, and tourism; decreased property values; and degradation of public water supply. Together, these drivers prompt regulatory agencies to develop nutrient management strategies that often require water resource recovery facilities (WRRFs) to implement some level of nutrient removal. Although there are many nutrient sources that affect watersheds, including agricultural practices, urban stormwater runoff, and the deposition of airborne pollutants, effluent from municipal wastewater treatment is a significant contributor in many urban and coastal areas

(Johnson Foundation, 2014). The level of nutrient removal required for a given municipality or facility is specific to the receiving water and is influenced by state laws and input from a variety of stakeholders. Finally, because treated wastewater is increasingly being used for beneficial reuse applications such as groundwater recharge and grey water supply, more stringent and more complicated nutrient regulations must be taken into consideration as utility managers develop a nutrient management plan.

1.1.1 Regulatory Requirements

Nutrient discharges into receiving waters originating from point sources such as WRRFs are generally regulated on a state or regional level. The Federal Clean Water Act (CWA) calls for waterbodies to be "swimmable and fishable" and allows for states and regional agencies to determine the regulatory mechanisms (i.e., total maximum daily loads [TMDLs], revised discharge permits, consent orders, etc.) to be used to control point-source nutrient discharges, but federal law does not require specific nutrient criteria. However, the U.S. Environmental Protection Agency (U.S. EPA) has authority for nutrient standards if regional or state agencies to not take appropriate actions (i.e., state of Florida numerical nutrient criteria). Regional (i.e., U.S. EPA regional offices) and state agencies are tasked with determining the need for nutrient management for specific watersheds and waterbodies, and have the authority to set, promulgate, and enforce specific nutrient limits. Generally, water quality regulations are becoming more stringent for many waterbodies in North America. The mechanism by which specific agencies develop and implement nutrient management regulations varies greatly based on state law and the specific uses and characteristics of the affected waterbody. The establishment of effluent nutrient limits for WRRFs can be influenced by many variables, including receiving water quality; the number and type of discharges affecting the waterbody; the hydraulics of the waterbody; the discharge flow characteristics (i.e., mixing zones and dilution factors); and downstream water uses, effects, and water quality requirements. Because each state may have a different process by which nutrient limits are set and promulgated and different timelines for meeting nutrient goals, municipalities should become familiar with the specific processes and timelines to effectively implement nutrient removal to meet new, proposed, or future discharge limits.

1.1.2 Watershed Water Quality and Total Maximum Daily Loads

Water quality standards provide the legal basis to control nutrient discharges and protect water quality. Watershed water quality may have multiple environmental, public health, and environmental effects, and nutrient criteria are applied to specific receiving waters in an effort to improve and/or maintain water quality. Most nutrient criteria are the basis for permit limits included in National Pollutant Discharge Elimination System and State Pollutant Discharge Elimination System permits. If a waterbody is already identified as impaired for nutrients, the numerical value of a nutrient limit in a specific permit is typically based on the TMDL set for the impaired receiving waterbody. A TMDL is a calculation of the maximum amount of a pollutant that a waterbody can receive and still meet water quality standards. The TMDL takes into account the allocation of that load among point sources (WRRFs) that receive a waste load allocation and nonpoint sources that receive a load allocation. In addition, TMDLs include (implicitly or explicitly) an additional load that is added on to the TMDL as a safety factor (U.S. EPA, http://water.epa.gov/lawsregs/lawsguidance/cwa/tmdl/overviewoftmdl.cfm). The allocation of the TMDL to a specific WRRF is determined by the flows and loads discharged by the WRRF and the level and degree of dilution and effect of that discharge on the quality of the receiving water. This information is typically used to establish nutrient removal requirements specific to the regulated utility. Most nutrient criteria are initially applied as TMDLs; however, as nutrient regulations evolve and become more stringent, the criteria can result in numerical limits that become defined effluent permit limits.

1.1.3 Effects of Nutrient Management Beyond the Watershed

Because of the broad effects of poor water quality, input from myriad stakeholders is needed to develop a cost-effective nutrient management plan for a given watershed. These stakeholders include the municipality and its customers, regulatory agencies, groups that use the watershed, and businesses that rely on watershed quality. Beyond the affected watershed, effective nutrient management may have significant social and economic effects on the surrounding communities. Improving water quality can increase or improve business activity associated with the watershed, including fishing, shellfish harvesting, and recreation. In addition, improved water quality can elevate land values, resulting in increased development and further economic growth. Finally, other stakeholders such as those associated with agriculture, energy, and any commercial or industrial entity that produce nutrients may be affected by a nutrient management plan.

A nutrient control program that includes advanced wastewater treatment may affect total energy use, greenhouse gas (GHG) production, and the total carbon footprint of the facility and its daily operations. Most advanced treatment technologies that are designed to meet more stringent nutrient effluent standards are energy-intensive and require additional chemicals (i.e., supplemental carbon, alkalinity, etc.). The added energy and chemical requirements will increase the carbon footprint of the treatment facility and its daily operations, thus increasing total GHG production. It is difficult to quantify the true cost or effect of nutrient removal technologies on carbon footprint and GHG production of a specific facility, but is has long been theorized that the improvement of water quality may be achieved at the expense of atmospheric effects. The true costs and benefits of nutrient removal are site-specific and must account for current and future regulations, emerging technologies, and environmental and socioeconomic effects within and beyond the watershed. One possible option is to combine nutrient management with in-facility nutrient recovery, which may provide additional revenue while helping meet nutrient requirements. Nutrient recovery has the potential to produce higher-value agricultural products that are more acceptable to the general public compared to traditional products (i.e., land application of biosolids). However, like advanced nutrient removal, implementing nutrient recovery processes will require upfront capital investment, additional operator training, and increased operation and maintenance (O&M) costs.

1.2 Developing a Vision for Staff

1.2.1 Setting Short- and Long-Term Goals

As nutrient removal becomes a reality for a utility, it is important to set both short- and long-term goals. Short-term goals include "low-hanging fruit", as discussed in Section 1.2.2.2. There needs to be an understanding of what the capabilities of the WRRF are and what would be needed to achieve a short-term goal, such as changing process control parameters, adding instrumentation, and so on. Short-term goals must be realistic and achievable. As short-term goals are accomplished and staff has a clearer understanding of potential permit limits, then longer-term goals need to be established that are realistic and achievable. There also needs to be a partnership between managers, operators, regulatory agencies, and financial institutions in setting goals, particularly long-term goals that most probably will have a significant cost associated with them.

1.2.2 Determining Treatment Level to Meet Goals

1.2.2.1 Low-Hanging Fruit: Low-Cost Modifications to Achieve Some Benefit

Existing activated sludge treatment facilities have the ability to remove nitrogen to some extent greater than the nitrogen removed by assimilation, depending on whether they have excess capacity or not. Often, treatment facilities not specifically designed for biological nutrient removal (BNR) can remove as much as 70% or more of the influent nitrogen by just modifying how they operate the process or through low-cost retrofits that include installing nitrate recycle pumps and creating anoxic zones. The vision for staff at a WRRF should be to be innovated and creative. The main concern and, therefore, barrier to innovation is violating discharge permits. There needs to be cooperation between regulators and facility operations personnel because environmental benefits can be achieved with simple process modifications. Additionally, using ammonia-based dissolved oxygen control and optimizing simultaneous nitrification and denitrification are other ways of achieving some level of nitrogen removal and will help to meet short-term goals.

1.2.2.2 Nutrient Removal Versus Nutrient Recovery

There are technologies (i.e., engineered stuvite recovery) for phosphorous recovery that have been proven at full scale. These technologies also remove some ammonia through the precipitation of magnesium ammonia-phosphate (i.e., struvite). The process does require that the WRRF has anaerobic digesters and, as such, is not necessarily applicable to many treatment facilities. There are no proven methods for ammonia recovery at this point in time, although such technologies are currently under development. However, land application of biosolids recovers not only nitrogen and phosphorous and trace elements, but also water. It is important for staff to understand the technologies, pros and cons, and costs and effects on other treatment processes so that an informed decision can be made for nutrient recovery. Certainly, the goal of all WRRFs should be to recover as much of the nutrients and energy as is possible and practical.

1.2.2.3 Understanding Effects on Energy Use and Budgets

Current nutrient removal processes are energy-intensive. There is a significant increase in energy use for facilities that nitrify, and additional chemical costs for high-level denitrification and phosphorous removal. Furthermore, there are increased O&M costs and additional laboratory testing requirements. These costs need to be included in budgets, and ratepayers must be educated to understand the reasons for these increases. It is imperative that WRRFs seek out the most cost-effective methods of nutrient removal and be willing to take some risk for newer, less-tested processes. For example, a number of deammonification technologies are currently available that effectively treat high-strength sidestreams. In addition, other technologies including simultaneous nitrification/denitrification and mainstream deammonification are currently being developed and demonstrated in North America. These technologies provide significant chemical and energy savings, and currently represent state-of-the-art processes for nutrient removal.

1.2.2.4 Associated Goals

There are a number of goals that municipalities can strive to achieve while pursuing compliance with current and new nutrient limits. These goals may include

- Maximizing the capture and reuse of waste stream nutrients,
- Minimizing the energy used to process wastewater,
- Further minimizing nutrient release into the environment,
- Minimizing alterations to the hydrologic cycle,
- Minimizing the release of GHG emissions from process operations,
- Maximizing economic benefits, and
- Minimizing complexity of process O&M.

Achieving some or all of these goals is the basis of the "facilities of the future" concept (see Chapter 1); moreover, these functional goals are quickly becoming the guiding principles of sustainable nutrient management.

1.3 Stakeholder Involvement

Without the understanding of ratepayers and other stakeholders, nutrient goals often cannot be met. There needs to be collaborative efforts between the WRRF, ratepayers, local watershed groups, legislators, regulatory agencies, and other nongovernmental organizations. Stakeholders must be educated that nutrients are affecting water quality, use of waterbodies, commercial and recreational activities, and, in some cases, the region's economy. Stakeholders need to be involved from the inception of a nutrient strategy.

2.0 STAFF EDUCATION AND DEVELOPMENT

2.1 Making Training a Requirement for All

As nutrient management becomes more of a reality for staff, they must have a thorough understanding of why it is required and how it is done with both current and emerging technologies. Training must include all staff, from utility managers to frontline operators, but must be conducted in the right sequence and with the best educators/educational material possible. When utilities are first faced with nutrient removal, education is the first step in allowing staff to understand and embrace this new requirement. Starting early is important and should include educational programs on why nutrient removal is required and the effect on the environment and implementation schedule. Understanding the "why" is critically important and must be done before understanding the theories and technologies behind nutrient management.

Furthermore, training needs to be accomplished in steps, from basic to accepted technology to emerging technology, thus empowering staff to implement nutrient removal in an effective and efficient way, but also to embrace advances in technology. Everyone involved in utility operations should be familiar with the publication, *The Water Resources Utility of the Future: A Blueprint for Action* (WEF et al., 2013). *The Nutrient Roadmap*, in part, is preparing for the "utility of the future".

2.1.1 Basic Understanding of Nutrient Removal Theory

Once staff understands the reasons behind the regulations, then training on how nutrient removal is accomplished needs to begin. Nutrient removal can include nitrogen removal, phosphorous removal, or both; additionally, even though a facility might only require management of one of these nutrients, it is important for staff to understand the technology and challenges for both.

Operations personnel must understand the basic theory and practice of nutrient removal including chemical precipitation of phosphorous, biological phosphorous removal, BNR, and the biological removal of both nutrients. They need to understand the effect of upstream processes on nutrient removal and the effect of nutrient removal on downstream processes. They also need to understand the biological process and how environmental conditions such as dissolved oxygen, pH, and temperature affect process performance. Furthermore, they need to understand how discharges to the WRRF may cause inhibition of the biological process and, therefore, the potential effect of industrial discharges to their facility.

2.1.2 Understanding Current Practices

In addition to training on nutrient removal theory, personnel should understand current nutrient removal processes and what level of nutrient removal they can achieve. There are several nitrogen removal processes that are commonly used, and having knowledge of the various process configurations can help in selecting what process is best for a facility given location and permit requirements; the same is true for phosphorous removal. If staff understands the limits that can be attained by chemical phosphorous removal, biological phosphorous removal, or combined chemical and biological removal, then they can make informed decisions on what is best for their facility. For example, if the five-stage BardenPho process is used to remove both nitrogen and phosphorus, operators must be trained to understand the importance of readily biodegradable chemical oxygen demand and its role in both enhanced biological phosphorus removal and dentrification so they can cost-effectively manage process performance to meet permit requirements.

Staff also needs to understand the effect on O&M costs, such as higher energy costs for nitrification, chemical use for denitrification, or higher solids disposal costs because of chemical phosphorous removal.

2.2 Ensuring High-Quality and Appropriate Training for Various Staff Levels

It is vitally important that staff receives the highest quality training at the appropriate level. Training should reflect the state of the art and represent the current best practices, process control techniques, and operations and maintenance requirements and costs. There are many good training documents, Web casts, courses, and seminars that can provide high-quality training that is appropriate for utility managers down to operations personnel, from basic to advanced levels. Water Environment Federation training materials such as *Operation of Nutrient Removal Facilities* (WEF, 2013) and *Nutrient Removal* (WEF, 2010) are examples of documents with current and peer-reviewed information on all nutrient removal processes, theory, management, and control.

Utility managers must make the effort to provide high-quality training and ensure that sufficient budgets are available for training. Networking and continuing education are critical to ensuring facility staff understand current and emerging technologies and are adequately prepared to meet their facility's current and future nutrient permit requirements. The only way a facility can proceed successfully along the road to efficient nutrient removal/recovery is by having highly trained staff.

2.3 Understanding Emerging Technologies and Their Potential Place in the Industry

Along with a thorough understanding of current technology and operation practices, personnel need to be educated about emerging technologies and what benefits they might have. Typically, emerging technology connotes risk and risk taking. Understanding how these technologies work, potential cost and resource savings, and potential resource recovery allow staff to make informed decisions on what is best for their facility, both to meet current and future needs and to determine what level of risk is necessary.

Permits may continually reduce the amount of nutrients that can be released. As such, future planning is critical, and understanding new technologies must be part of that planning process. Not all technologies are the right fit for a given facility. Some require a higher skill level or more sophisticated control techniques. A thorough understanding of what is right for a particular WRRF requires knowledge of all available and potential technology.

2.4 Empowering Staff

2.4.1 *Allowing Staff to Make Changes to Meet Visionary Goals*

Setting visionary goals is an important step in the nutrient removal roadmap. The visionary goals must not be set only by managers, but must be a collaborative effort by various levels and disciplines of staff. The next step is to empower the staff to make changes to meet these goals. This is not an easy thing for many managers because it involves taking risks and allowing staff to have more responsibility. However, by doing this, staff will take ownership of the process, which typically results in a more successful operation. The visionary goals should be reviewed frequently by all staff members, and the status of achieving those goals should be discussed. Operations personnel may have ideas that allow higher levels of nutrient removal with only simple modifications, or they may have ideas on operations or maintenance. To be successful, everyone must feel that they are an important part of the process.

2.4.2 *Allowing Staff to Be Part of the Decision-Making Process*

As decisions are made as to what process or technology to install at the treatment facility, all levels of staff and operations personnel must be involved in decision-making. Each level will bring questions, ideas, or concerns that perhaps another level never considered. In addition, when all staff is involved in the decision, when the process has been installed, there will be a greater effort to ensure that it meets expectations and is operated and maintained at the highest quality standard.

3.0 FINANCING

3.1 Developing a Short- and Long-Term Strategy

Financing and budgets play an important role in developing a strategy for nutrient removal/recovery. Just as with staff development, training, and evaluating technology, a plan must be developed to progress from no removal/recovery to advanced stages in which nutrients are recovered in some form, whether it is through land application or irrigation or through product development.

3.2 Identifying the Costs

A significant part of the strategy is to understand what category of costs needs to be planned. There may be significant capital costs, an increase in O&M costs, and costs for training and staff development, but there might also be cost avoidance, which will help to repay some of the debt. Other considerations are existing facility size and complexity. When developing budgets for a nutrient management plan, in addition to capital and standard O&M costs (i.e., salaries and benefits), the facility must also consider costs associated with new equipment, equipment replacement and repair, and laboratory needs specifically related to nutrient removal operations. A nutrient strategy must be realistic and may not be applicable at a small WRRF without an extremely high cost per pound of nutrient removed or recovered.

3.2.1 Capital Costs

All nutrient removal/recovery processes require some level of capital investment. Some nutrient removal can be accomplished simply through process changes or low-cost retrofits, and these might be the first steps for a WRRF. For higher levels of nutrient removal and for nutrient recovery, capital costs can be significant and require planning, coordination with funding agencies, and public education. As a nutrient strategy is developed and agreed upon, a capital improvement plan must be integrated to that strategy, including a time schedule and potential funding sources. Without a viable and well-developed capital plan, no matter how good and well thought out a strategy is, it will not succeed.

3.2.2 Operating Costs

Typically, nutrient removal/recovery processes require more staff, including operators, laboratory personnel, instrument technicians, and maintenance workers. Some of the duties can be outsourced, such as instrumentation and maintenance; whether the process is done inhouse or is outsourced, however, operating costs will increase. Furthermore, there may be additional costs for chemicals such as external carbon sources or chemicals to adjust alkalinity. However, there are also opportunities for cost savings, especially in electrical energy. A utility manager needs to understand the expected increase in operating costs, develop a plan for educating ratepayers on the need for any potential increases and why nutrient removal/recovery is required, and plan budgets over longer terms. Using life cycle analysis is crucial in determining the process that will best serve the needs of the community and meet regulatory requirements.

3.2.3 *Staff Development*

As mentioned in Section 2.0, staff needs to be trained, beginning with an understanding of the effect of nutrients on water and ecosystem quality; the basic theory and practice of biological nutrient removal; and more advanced technical, theoretical, and process control requirements needed to maintain consistent regulatory compliance. Furthermore, additional staff may be required with different skill sets. The utility manager must evaluate what training is needed, staff discipline, how the training will be done, where it will be done, and how long a period it will take to develop the staff to meet the nutrient strategy and vision developed by the organization. Along with a training program, the budget must be adequate to meet all staff development needs, which may require long-term budget planning.

3.3 Identifying Sources

Identifying funding sources may be one of the most complex aspects of a good nutrient strategy and is probably the most important step to a successful program. Funding sources may include typical types such as loans and grants, but may also come from technology companies that are looking for a facility to demonstrate their process. The goal of identifying funding sources is to look at all possibilities, even those that have never been considered.

3.3.1 *Loans*

Many states provide low-cost loans for nutrient removal/recovery. These are typically part of the state revolving fund (SRF) loan program. However, SRF programs in many states are stressed and may only provide a small percentage of the total cost of a project.

3.3.2 *Grants*

Grants may be even more difficult to obtain, at least from state agencies. The amount of grant money available tends to be small, and there is significant competition for grants. There may be nontraditional agencies that are willing to give grants for nutrient removal/recovery, such as the U.S. Department of Agriculture or the U.S. Department of Energy, if they provide for nutrient recovery through fertilizer or irrigation or energy reduction. It is important for the utility manager to continually search for grants, especially as the utility progresses from basic to more complex and intense methods of nutrient recovery.

3.3.3 *User Rates*

Certainly, user rates can be increased to cover the cost of nutrient removal/recovery projects; however, requests for increases must be coupled with an intense and well-designed public education program. Ratepayers, agency board members, and other decision-makers need to understand why a project is required, what the regulatory drivers are, the potential penalties if a project cannot be financed, and the long-term plan to meet the vision of nutrient removal/recovery. These same entities need to understand how rates will increase currently and in the future. They also need to understand the potential for revenue from a product sale and how that might reduce user rates. Changes in user rates need to be coupled to the capital improvement program and the long-term nutrient strategy.

3.4 Identifying Alternate Financing Opportunities

There is a potential for funding from sources other than those typically considered. Just like evaluating nutrient processes, identifying alternate financing opportunities is just as important or perhaps more important.

3.4.1 *Private Versus Public Finance and Operations*

Technology providers often are looking for WRRFs that are willing to host a demonstration project. These technology providers may fund a project in its entirety or may share costs with the utility. In the case of a technology that provides a product, the provider may use the revenue stream to fund the project. Having a demonstration project that potential users can see offers a great value, and can be a way in which a utility can meet its visionary goals. There are some potential barriers to this kind of financing, such as local regulations or public perception; as such, planning and education are, again, critical to using private financing. Changing regulations and changing public perception all take time and need to be factored into the strategic nutrient plan implementation schedule.

3.4.2 *Alternative Project Delivery*

There are many potential alternate project delivery plans that can be used. A good discussion of various project delivery methods is provided in *Upgrading and Retrofitting Wastewater Treatment Plants* (WEF, 2005). Each delivery method has potential benefits and needs to be evaluated based on local regulations, existing union contracts, and local ordinances.

3.5 Identifying Revenue Streams

As plants progress toward resource recovery facilities, the potential for revenue streams increases. This potential must be evaluated as different technology is being investigated. For instance, a facility may consider the sale of GHG credits for nitrous oxide (N_2O) emission reduction and renewable energy generation.

3.5.1 Nutrient Recovery and Sale of Materials

The sale of biosolids as fertilizer has been practiced for many years and is certainly a successful way of generating revenue while recovering and reusing the nutrient value of biosolids. Fertilizer can be made through anaerobic or aerobic digestion, heat drying, composting or alkaline stabilization, or a combination of these. In addition to nutrients, land application of thickened or dewatered biosolids can provide water to crops, and treatment facility effluent can be used to irrigate orchards or other crops, providing water and nutrients.

3.5.2 Nutrient Credit Trading Programs

There have been several successful nutrient trading programs. These programs have several benefits for the WRRF and ratepayers. Trading programs can create incentives for the municipality or utility to move forward with nutrient removal at a faster rate than is typical because selling of nitrogen credits could offset some of the costs associated with upgrading to nutrient removal. For those states that do not have existing programs, there are opportunities for them and stakeholders to examine trading programs from other states or regions to see which best suit their needs. Utilities, regulators, and stakeholders can get a good understanding of various trading programs in the publication, *Water-Quality Trading: A Guide to the Wastewater Community* (Jones et al., 2005).

4.0 PLANNING AND PRIORITIZATION TO ACHIEVE COMPLIANCE

4.1 Planning to Meet Current and Future Regulatory Requirements

Some regulatory agencies provide notification and guidance of pending and future regulatory changes either within the utility's actual discharge permit or through other communications. One of the primary challenges for utility managers is being able to effectively evaluate and plan for both short- and long-term nutrient criteria in a regulatory and economic environment that is increasingly multifaceted and dynamic. Population growth, aging infrastructure, increasingly complex water quality issues, and the dynamic economic environment make is especially challenging for utilities to cost-effectively meet current and future regulatory requirements. Utility managers need to engage all stakeholders to properly define the specific drivers that control nutrient requirements and determine what steps can be taken to successfully achieve compliance with new and pending requirements. This can be especially difficult if the actions needed to meet specific nutrient criteria interact or hinder the ability to meet other effluent criteria or visa versa (i.e., effect of low ammonia on maintaining total residual chlorine limits).

Strategies to achieve regulatory compliance may focus on controlling point sources or a combination of point and nonpoint sources. Utilities may benefit from considering approaches that are watershed or system based, such as water quality trading and integrated planning. Integrated planning and management approaches may be applied to help utilities achieve simultaneous compliance with multiple, and possibly conflicting, effluent criteria (U.S. EPA, 2012). Utilities must consider all criteria, including nutrient criteria, and how seemingly different criteria may interact and possibly affect a facility's ability to simultaneously meet all of their permit criteria in a cost-effective and consistent manner.

In 2012, U.S. EPA developed an *Integrated Municipal Stormwater and Wastewater Planning Approach Framework* for utilities (U.S. EPA, 2012). An integrated planning process has the potential to identify a prioritized critical path to simultaneously achieve multiple water quality objectives by identifying interactions efficiencies in implementing competing requirements that arise from different water quality criteria. These issues can be magnified if municipalities are faced with competing wastewater and stormwater water quality projects, including capital investments and O&M requirements. An integrated planning approach may lead to more sustainable and comprehensive solutions, such as green infrastructure, that improves water quality and supports other indirect economic and social benefits resulting from improved water quality. Federal, state, and local regulatory agencies must implement regulations, policy, and guidance that provide the necessary flexibility to implement an integrated planning process.

An integrated planning approach is not about lowering existing regulatory or permitting standards or delaying necessary improvements. Rather, it is intended to be an option to help municipalities meet their multiple water quality obligations by optimizing the use of their existing infrastructure; it is also intended to help municipalities make the most prudent use of their capital investments through the prioritization of regulatory drivers and the appropriate sequencing of work based on scientifically supported data and within the existing regulatory process.

4.2 Water Quality Trading

Water quality trading is a market-based approach to improve and preserve water quality. In 2003, the U.S. EPA issued a *Final Water Quality Trading Policy* to provide guidance to states and tribes on how trading can occur under the Clean Water Act and its implementing regulations. Trading can provide greater efficiency in achieving water quality goals in watersheds by allowing one source to meet its regulatory obligations by using pollutant reductions created by another source that has lower pollution control costs. U.S. EPA's policy endorses trading as an economic incentive for voluntary pollutant reductions from point and nonpoint sources of pollution and as a way to achieve ancillary environmental benefits such as creation of habitat (U.S. EPA, 2003).

While water quality trading has not been extensively implemented in the United States, nutrient trading has been successfully implemented in watersheds across the country. Nutrients are the most common water quality constituents in existing trading programs. In general, water quality trading approaches include point-to-point source trades, non-point-to-point trades, and a combination of the two approaches. For nutrients, as an example, a non-point-to-point source trade may consist of agricultural best management practices that achieved nutrient reduction that are used to offset nutrient loads in the effluent from WRRFs. U.S. EPA and others have provided guidance and information on water quality trading. The U.S. EPA Water Quality Trading Web page (U.S. EPA, 2014) includes information and links to numerous resources. Case studies of existing trading programs are useful for understanding various trading approaches, details of the trading agreements and credits, the process for implementing trading, and costs. The Electric Power Research Institute (2013), Branosky et al. (2011), and Morgan and Wolverton (2005) provide summaries of existing and planned trading programs. A summary of some of the nutrient trading programs currently in place is provided in Table 2.1.

4.3 Effects of Nutrient Management Beyond the Watershed

The true benefits and costs of nutrient management within and beyond the affected watershed are difficult to quantify. The primary purpose of nutrient limits is to protect public health and aquatic life. The multiple benefits of improved water quality within the watershed are well documented. There are also benefits beyond the watershed. If nutrient removal improves the

TABLE 2.1 Examples of existing water quality trading programs (adapted from EPRI [2013]).

Name/location	Constituent	Type of trade	Description
State of Connecticut	Nitrogen	Point source to Point source	Multiple point-source facilities participate in this program. Nitrogen credits are traded through the State of Connecticut's Nitrogen Credit Exchange
Southern Minnesota Beet Sugar Cooperative	Phosphorus	Point source to nonpoint source	Trading between the Southern Minnesota Beet Sugar Cooperative and beet sugar growers in the cooperative
State of Virginia	Nitrogen and phosphorus	Point source to point source and nonpoint source to point source	Multiple facilities participate in this program. Credits are traded through the Virginia Nutrient Credit Exchange.

overall water quality of a waterbody like the Chesapeake Bay, San Francisco Bay, or Long Island Sound, then this may allow the return of migratory birds and fish populations, improving not only the health of the specific watershed, but also relieving pressures on other watersheds, reducing the segregation of breeding grounds, and potentially providing additional water quality and economic benefits.

Municipalities and their stakeholders must consider the potential environmental effects, operational effectiveness, community effects, and economic factors associated with advanced nutrient removal or else they may trade one problem for another when developing solutions to remove nutrients from waste streams. The potential costs and tradeoffs of nutrient removal include

- Use of limited funding that might take away from competing priorities, including aging infrastructure that is not directly related to water quality or nutrient removal;
- Increased demand on the energy grid required to meet nutrient criteria (aeration);

- Increased carbon footprint associated with O&M of nutrient removal technologies, including additional chemicals (alkalinity and supplemental carbon) and additional biosolids disposal (processing and transportation);
- Increased GHG production via the production of nitrous oxides during nitrification and denitrification (Chandran et al., 2011); and
- Increased O&M costs independent of carbon footprint.

5.0 ESTABLISHING ASPIRATIONAL GOALS AND SUBGOALS

As municipalities strive to balance the economic, environmental, and regulatory pressures associated with meeting current and future nutrient criteria, they may want to consider specific goals and subgoals for achieving success. Some municipalities may also have aspirational goals that reflect their long-term goals for facility performance and operational sustainability. Treatment for nutrient removal/recovery can both address a critical need and present a real opportunity for environmental and economic gain, and should be developed in manner that is consistent with municipalities' short- and long-term goals.

5.1 Nitrogen and Phosphorus Recovery

With the increased energy demands and environmental effects of advanced nutrient removal processes, new technologies are being developed to improve the effectiveness and efficiency of advanced nutrient removal and recovery systems. Given the dual challenges of increased nutrient loads at wastewater facilities and the growing need to preserve and recycle these nutrients, many members of the wastewater sector believe the time is right for a sector-wide shift away from basic treatment and toward nutrient recovery and removal. Such a shift would complement conventional treatment and disposal and would harvest valuable resources from municipal waste streams.

As society is demanding cleaner water, new opportunities are being developed to remove nutrients from wastewater more effectively, and to recover them as resources for economic benefit. Although many nutrients can be costly or energy-intensive to remove, they can also be revenue-generating or at least cost-offsetting when recovered as a marketable resource. Recovering nutrient products (fertilizers) from municipal waste streams could help to reduce nutrient imbalances in natural water systems, produce a marketable product, and, in the case of phosphorus, provide a sustainable source of a finite resource to support agricultural production and food security in the United States and worldwide (Reijinders, 2014; Johnson Foundation, 2014). New and existing technologies are available for engineered struvite harvesting from phosphorus-rich digested sludge dewatering streams. Full-scale engineered recovery of struvite, a valuable slow-release fertilizer, using processes like Ostara® and Mulitform Harvest® has been used in Europe and at multiple locations in the United States. Additionally, new technologies are being developed to efficiently remove ammonia- and nitrogen-based fertilizer compounds from high-strength waste streams. It is believed that as nutrient regulations become more stringent and utilities seek new revenue sources to support operations, resource recovery processes such as struvite recovery will become more prominent.

Technological advances continue to drive progress toward smarter nutrient management. A wide array of nitrogen- and phosphorus-removal tools exist for WRRFs across a spectrum of sophistication and capital costs. These tools range from nitrogen management techniques for upstream sources (e.g., urine separation, water conservation, pretreatment, and protein recovery) to various facility-scale, treat-and-control approaches (e.g., conventional nitrification/denitrification, nitritation/denitritation, or nitrite shunt and deammonification or partial nitritation/anaerobic ammonium oxidation). If nutrient recovery is not applicable at a specific facility, then new and emerging nutrient removal technologies provide opportunities for municipalities to achieve compliance with current and future nutrient limits using less energy and chemicals. These technologies have the potential to greatly reduce operational costs of BNR, and will likely develop to the point where they are applicable at facilities of all sizes.

5.2 Energy and Water Recovery

In addition to nutrients, energy and water are two other resources that can be recovered or produced at WRRFs. Theoretically, the chemical energy contained in typical municipal wastewater is 5 times greater than the energy required to treat wastewater using traditional aerobic methods (Heidrich et al., 2011). The most developed and applied method for harnessing this inherent energy is the production of biogas via anaerobic digestion. There are a number of different digestion technologies that can be used to enhance biogas production. In addition, there are many pretreatment technologies that are being developed and demonstrated that may improve gas production and decrease total solids production. Finally, co-digestion using source-separated organic wastes such as food wastes, fats, oils, and grease, and local organic industrial wastes (i.e., yogurt whey) has been used at many municipalities to significantly increase biogas production to the point where the facility has achieved energy neutrality and, in some instances, has become a net energy producer. Achieving energy neutrality is likely too ambitious and costly for many municipalities, but maximizing the utilization of biogas, optimizing aeration efficiency, and implementing combined heat and power systems can be relatively low-cost improvements that can significantly reduce total energy costs.

Finally, the production of "purpose"-specific water for reuse is growing in popularity as specific regions of the United States and the world face scarcity of fresh water. Examples of purpose-specific water reuse include direct groundwater recharge to maintain groundwater supplies, industrial water reuse for cooling and boiler water, grey water reuse in dual-plumbing systems, irrigation water for non-edible crops, landscaping, golf courses, and direct potable water reuse. The specific type of reuse that might be considered by a municipality will be site-specific and will likely require upgrades to existing treatment systems.

For a municipality to consider implementing energy or water recovery, they must undertake a thorough cost–benefit analysis. This analysis must include the evaluation of new and emerging technologies and a determination that these technologies can be operated in conjunction with current nutrient removal process to ensure nutrient limits can be met. The application of advanced nutrient, energy, and water recovery is quickly becoming the goal of many large utilities. Future technology advancements will likely allow facilities of all sizes to transform into WRRFs that can simultaneously meet stringent nutrient limits while recovering valuable resources in the form of energy, biogas, reusable water, nutrients, and biosolids.

6.0 BECOMING CLIMATE READY

Climate change issues are important for WRRFs and utility managers. After recent significant storms such as Hurricane Irene in 2011 and Superstorm Sandy in 2012 on the East Coast, resiliency in design is critical as is protecting infrastructure from storm damage and surges. Wastewater facilities are estimated to be in the top 10 industries for emissions of GHGs. Some studies suggest that nitrogen removal might result in the emission of N_2O, which is a GHG; however, currently it does not appear to be significant. Still, as more WRRFs convert to BNR, it may become an issue and, as such, technology needs to be developed to reduce or capture any GHGs produced. As WRRFs develop their roadmap and goals for nutrient removal, they need to evaluate any known direct or indirect emissions of GHGs.

Additionally, WRRFs must ensure that they are resilient enough so that processes and equipment are protected. Severe environmental damage can result from storm-related damage and a considerable amount of time is needed to put these facilities back in operation. Water resource recovery facilities need to set goals to reduce any GHGs, prevent new sources, and ensure the facility is resilient.

7.0 REFERENCES

Branosky, E.; Jones, C.; Selman, M. (2011) *Comparison Tables of State Nutrient Trading Programs in the Chesapeake Bay Watershed;* WRI Fact Sheet Version 1; World Resources Institute: Washington, D.C.

Chandran, K.; Stein, L.; Klotz, M.; van Loosdrecht, M. (2011) Nitrous Oxide Production by Lithotrophic Ammonia-Oxidizing Bacteria and Implications for Engineered Nitrogen-Removal Systems. *Biochem. Soc. Trans.,* **39** (6), 1832–7.

Electric Power Research Institute (2013) *Case Studies of Water Quality Trading Being Used for Compliance with National Pollutant Discharge Elimination System Permit Limits*; Electric Power Research Institute: Palo Alto, California.

Heidrich, E. S.; Curtis T. P.; Dolfing, J. (2011) Determination of the Internal Chemical Energy of Wastewater. *Environ. Sci. Technol.,* **45**, 827.

Jones, C.; Bacon, L.; Kieser, M.; Sheridan, D. (2005) *Water-Quality Trading: A Guide to the Wastewater Community*; Water Environment Federation: Alexandria, Virginia.

Morgan, C.; Wolverton, A. (2005) *Water Quality Trading in the United States;* National Center for Environmental Economics Working Paper No. 05-07; U.S. Environmental Protection Agency: Washington, D.C.

Reijnders, L. (2014) Phosphorus Resources, Their Depletion and Conservation, a Review. *Resour. Conserv. Recycling,* **93**, 32–49.

The Johnson Foundation at Wingspread (2014) The Road Toward Smarter Nutrient Management in Municipal Water Treatment. The Johnson Foundation at Wingspread: Racine, Wisconsin; Mar. http://bit.ly/smarternutrient (accessed Mar 2015).

U.S. Environmental Protection Agency (2003) *Final Water Quality Trading Policy.* http://water.epa.gov/type/watersheds/trading/finalpolicy2003.cfm (accessed Feb 2015).

U.S. Environmental Protection Agency (2012) *Integrated Municipal Stormwater and Wastewater Planning Approach Framework*; U.S. Environmental Protection Agency: Washington, D.C.

U.S. Environmental Protection Agency (2014) *Water Quality Trading.* http://water.epa.gov/type/watersheds/trading.cfm. June 05, 2014. U.S. Environmental Protection Agency: Washington, D.C.

Water Environment Federation (2005) *Upgrading and Retrofitting Wastewater Treatment Plants;* WEF Manual of Practice No. 28; McGraw-Hill: New York.

Water Environment Federation (2010) *Nutrient Removal*; WEF Manual of Practice No. 34; McGraw-Hill: New York.

Water Environment Federation (2013) *Operation of Nutrient Removal Facilities*; WEF Manual of Practice No. 37; Water Environment Federation: Alexandria, Virginia.

Water Environment Federation; Water Environment Research Foundation; National Associaton of Clean Water Agencies (2013) *The Water Resources Utility of the Future: A Blueprint for Action;* Water Environment Research Foundation: Alexandria, Virginia.

3

Understanding the Environment

1.0 INTRODUCTION

Wastewater reclamation utility managers are faced with reducing nitrogen and phosphorus discharges to a greater degree than that provided by current secondary treatment processes (WEF, 2010). Effluent limits vary widely among utilities now and may be subject to change in the future. This is because the response to nutrient discharges is not the same for different receiving waters. States are at different points in developing nutrient management strategies and adopting nutrient criteria. Therefore, the regulatory requirements vary among the states and may change in the future. Unlike many constituents subject to effluent limits, nutrients are not toxic at typical concentrations in secondary effluent (except un-ionized ammonia, which is a component of total nitrogen discharged). Rather, elevated nutrient concentrations may cause or contribute to increased algae, including those secreting toxic compounds, and aquatic macrophyte growth in the immediate vicinity of the utility or contribute to hypoxic conditions at a great distance from the discharge. Other sources of nutrients, including agricultural runoff and drainage and urban stormwater, contribute to, or may be the predominant source of, the nutrient loading to a receiving water. There are multiple additional factors that contribute to these eutrophic conditions, including waterbody morphology, stream flow, impoundments, available light, substrate, habitat, water color, and temperature.

Each utility manager has differing capabilities to remove and recover nutrients at their facilities and differing spatial opportunities to add or modify processes to increase removal. In addition to capital costs and ongoing increased labor operational expenses, removal processes may require significantly more energy and chemical use and result in increased direct or indirect greenhouse gas (GHG) emissions that should be considered.

Additional nutrient removal may affect biosolids production and alter biosolids reuse potential.

This section is intended to help utility managers identify the information needed, the tools to provide accurate data and analysis, and the factors to be considered to make an informed decision for long-term cost-effective and beneficial nutrient recovery or removal at their facilities.

2.0 EXISTING AND POTENTIAL FUTURE REGULATORY REQUIREMENTS

The Clean Water Act (CWA) authorizes states and sovereign tribes to establish water quality criteria for numerous elements and compounds that support designated uses for waters of the United States. States must establish water quality criteria that are at least as protective of uses as U.S. Environmental Protection Agency (U.S. EPA)-recommended criteria and must consider protection for downstream uses (40 CFR 131.4). In 1998, U.S. EPA proposed comprehensive criteria for total phosphorus and total nitrogen based on ecoregions. Ecoregions are large-scale ecological and geographic areas of similar climatic, geologic, soil, and hydrologic characteristics. The recommended ecoregion criteria are generally low concentrations; ecoregion criteria have not been confirmed by most state statistical evaluations comparing nitrogen and phosphorus concentrations with biological indices in their waters. States have also found that relationships between nutrient concentrations and biological indices can be "fuzzy" and difficult to discern. Few states have adopted numeric nitrogen and phosphorus criteria for all waters. Numeric nutrient criteria are more common for lakes and impoundments and less so for free-flowing streams.

States are required to review water quality data triennially. In addition, states must assess waters for compliance with water quality criteria on a biennial basis in even-numbered years. Where water quality criteria are not met, that segment is listed as impaired, and placed on a state's 303(d) list. The lists must be approved or amended by U.S. EPA. Impaired waters are then subject to a total maximum daily (TMDL) determination (40 CFR 131.4). The TMDL study identifies what loading reductions are required from each source in the impaired segment. If numeric nutrient criteria have been promulgated, effluent limits for nutrients will likely be established for the water resource recovery facility (WRRF).

The current general requirements for wastewater treatment in this country are illustrated in Figure 3.1. The CWA (PL 92-500 Section 301) mandated secondary treatment beginning in 1977. Typically, secondary treatment is considered to be 85% removal of 5-day biochemical oxygen

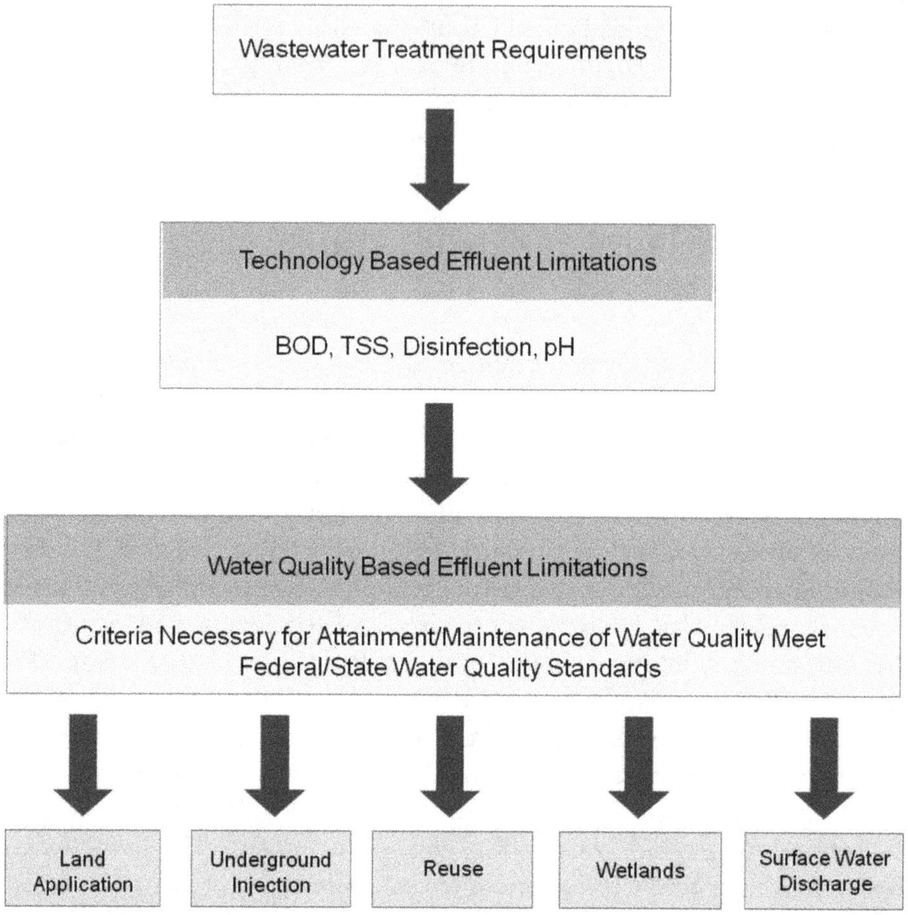

FIGURE 3.1 General requirements for wastewater treatment.

demand (BOD5) and total suspended solids (TSS) and a monthly average of 30 mg/L of BOD5 and TSS. Secondary treatment also requires an effluent with a nearly neutral pH and adequate disinfection as measured by the total coliform concentration. The aforementioned are commonly referred to as *technology-based effluent limitations.*

In locations where secondary treatment is not adequate to maintain public health or water quality standards, the state or U.S. EPA may mandate a higher level of treatment. These water-quality-based effluent limitations (WQBELs) can be required for continued discharge of treated effluent (or *reclaimed water*, as it is more aptly called) to any number of reuse or disposal methods. The most common WQBELs involve limitations on the discharge of phosphorus or forms of nitrogen. It is generally safe to assume that nearly all facilities that discharge to surface and groundwaters will eventually be required to implement some form of nutrient control. The choice often becomes one of meeting probable future regulations now, or providing the flexibility to upgrade later.

If the numeric nutrient criteria are not met, a WRRF will have National Pollutant Discharge Elimination System (NPDES) effluent limits, which are determined through a wasteload allocation calculation established in a TMDL. Water quality criteria for nitrogen and phosphorus, where promulgated, are at concentrations significantly lower than that of typical secondary effluent. Total maximum daily loads also identify reduction needed from urban-point and agricultural or other nonpoint sources. Reductions are generally achieved through best management practices (BMPs) that are largely unmonitored. Again, for most states with nutrient water quality criteria, the criteria are developed specific to the type of waterbody (e.g., small streams, large streams, lakes, and estuaries). It should be noted that water quality data for impairment determination are often limited and, therefore, it may be beneficial for a utility to supplement the data set to ensure that the watershed segment is correctly classified as impaired. Development of site-specific standards is also allowed in some states to reflect unique conditions not recognized in the general classifications. Site-specific criteria are most often proposed by WRRF utilities to more accurately characterize their receiving water and the subsequent waste load allocation.

While few states have promulgated nutrient criteria, most states have narrative standards that prohibit nuisance conditions. For nutrients, nuisance conditions are generally defined as excessive algae growth or observed excessive dissolved oxygen and pH swings in the receiving waters. In these cases, the state agency will assess the downstream conditions and set a nutrient effluent limit meant to correct the nuisance conditions. It is important to note that when streams are at low-flow/drought conditions, the WRRF may constitute a significant portion of stream flow, providing a continuous source of nutrients that induce nuisance conditions.

Dischargers in certain large regional watersheds need to consider an additional nutrient reduction requirement. For the Great Lakes watershed, all public NPDES permit holders have been subject to a 1.0-mg/L total phosphorus technology-based effluent limit. For the Mississippi River watershed, each state has been asked for a plan to reduce nitrogen and phosphorus loadings from their state by 45%, with the goal of reducing the extent of the hypoxic zone in the Gulf of Mexico. This may translate into a nutrient reduction requirement for some WRRFs in the watershed. In some states, the management plans propose WRRFs to reduce effluent total phosphorus to approximately 1.0 mg/L and total nitrogen to approximately 8.0 to 10.0 mg/L (i.e., a technology limit rather than a water-quality-based effluent limit). Phosphorus is not toxic at concentrations in typical WRRF effluent; rather, it induces excessive algal growth. For this reason, permit compliance may be based on monthly, seasonal, or annual averages. Nitrogen may also induce algal growth, but, depending on the nitrogen form and designated use of the receiving water, may be subject to limits based on toxicity as well. U.S. EPA (2013) recently published updated ammonia water quality criteria. If receiving water is designated for drinking water use, there is a 10-mg/L nitrate/nitrogen maximum contaminant level for drinking water (http://water.epa.gov/drink/contaminants/basicinformation/nitrate.cfm). Nitrate and nitrite toxicity to aquatic organisms has been studied (U.S. EPA, 2010). Studies have proposed nitrate/nitrogen aquatic life criteria (Camargo et al., 2005; Hickey and Martin, 2009;). Although U.S. EPA is not currently proposing aquatic life nitrate/nitrogen criteria, WRRFs should be aware of any future proposals. Water resource recovery facilities should also be aware of and utilize all permitting flexibility with respect to their nutrient effluent limits.

Water reuse may be considered as an alternative to treatment to extremely low concentrations if a user can be found that does not require low concentrations of total nitrogen and phosphorus. Water resource recovery facilities should also be aware of other effluent limits (e.g., chloride or total dissolved solids) that may be compromised by chemical addition to remove phosphorus. Regions of water shortage are exploring "one water" management, which integrates overall water supply and opportunities for water reuse. In cases where water is reused for agricultural or residential irrigation, nutrient removal may be unnecessary or the presence of nutrients may be advantageous.

As an alternative to removal of nitrogen and phosphorus to low levels solely at the WWRF, many states have put in place watershed programs that allow collaboration with other stakeholders to achieve overall water quality goals. Nutrient trading occurs where the WRRF pays for reductions at another point or nonpoint source with a lesser reduction at the WRRF (dnr.wi.gov/topic/SurfaceWater/WaterQualityTrading.html) (CBC, 2012; WRI, 2013).

Trading programs may theoretically achieve the watershed water quality goals at a lesser overall cost; however, this has yet to be proven on a large scale. Adaptive management is a structured iterative process of decision-making that recognizes the uncertain response of complex systems to specific actions. With respect to nutrient management, multiple stakeholders are organized within a watershed to each make nutrient incremental loading reductions through treatment and BMPs. Adaptive management requires a robust instream monitoring program conducted long term to capture varying hydrologic conditions to demonstrate that all components are successfully making progress toward achieving water quality goals (www.madsewer.org/Programs-Inititatives/Yahara-WINs) (Camargo et al., 2005). Subsequent actions are based on monitored results and evaluations to identify effective additional nutrient reduction practices.

Water resource recovery facilities that are required to treat to low concentrations of nitrogen and phosphorus may be faced with costs that would be a burden to the community. Most states have local or statewide variance procedures for relief, either in the ultimate treatment requirements or time for compliance. State variances must conform to and be approved by U.S. EPA requirements. Variances are time-dependent and must be re-evaluated periodically.

In summary, there are multiple state and federal regulations that will control the nutrient reductions required at a utility manager's WRRF. Identification of these regulatory requirements and compliance options at the beginning of the planning process is necessary to understand the data needs and analytical tools appropriate to make informed decisions for nutrient recovery and removal.

3.0 WATER QUALITY RESPONSE TO NUTRIENT LOADING—UNDERSTANDING THE RECEIVING WATER ENVIRONMENT

The CWA goals of attaining fishable and swimmable uses in our nation's lakes, streams, and estuaries are the primary driver for reductions in nutrient loads. The response of a waterbody to phosphorus and nitrogen loadings is site-specific and may be complex. Systems that have similar nutrient concentrations may display different symptoms of eutrophication. Geography, hydrology, hydraulics, and ecology are key factors in determining how the system will respond. Therefore, a one-size-fits-all standard or solution for setting nutrient load targets and allocations for specific sources will not achieve a consistent response.

Understanding the site-specific load-response characteristics for an individual waterbody is critical and can be complex. This understanding can be used to establish nutrient load targets that are based on the best available science and, once achieved, will result in attainment of the desired water quality response. The basic steps in developing this understanding include the following:

- Develop a conceptual model of the system,
- Select a modeling approach for assessing the load-response relationship,
- Identify data needs and compile and assess required data, and
- Develop nutrient load targets.

Because the load response is best evaluated over a larger watershed area, a utility may be best served by participating in water quality modeling conducted by a watershed group, a water quality management agency, or the state (typically for a TMDL), ensuring that the model will address the questions to be answered for the utility. The primary benefits of, and key considerations for, each of these steps will be briefly discussed in this section.

3.1 Site Conceptual Model

A site-specific conceptual model is a key first step in understanding the load-response relationship. A conceptual model describes the most important aspects of the system relevant to the problem of interest. Development of a site-specific conceptual model begins with defining the endpoints of concern or the environmental response indicators, then identifying the sources of nutrients, and, finally, linking cause and effect by identifying the key attributes of the physical, chemical, and biological system at play.

3.1.1 Environmental Response Endpoints of Concern Related to Nutrient Loads

Identifying the endpoints of concern, or environmental indicators, determines the level and complexity of all subsequent steps in understanding water quality response to nutrient loads. The typical response endpoints of concern related to nutrient inputs include

- Nuisance algal growth affecting aesthetics or recreational activities/ excess biomass;
- Harmful algal blooms releasing toxins or ammonia toxicity;
- Excess algal growth resulting in hypoxic conditions (depressed dissolved oxygen), decreased water clarity, or degraded habitat and fish or benthic macroinvertebrate community; and
- Taste and odor issues requiring additional treatment and resources by public water suppliers.

The endpoint of concern may be an issue locally or on a larger scale, such as hypoxia in the Chesapeake Bay or Gulf of Mexico.

3.1.2 Sources of Nutrient Loads

An assessment of all potential sources of nutrients to the waterbody of concern needs to be conducted and an inventory developed. Typical sources of nutrients include WRRFs (municipal, industrial pretreatment, water supply additives, etc.); combined sewer overflows; urban stormwater;

agriculture (cropland and feedlots); direct industrial discharges (treated process discharges and cooling water); erosion (upland, gully, and stream-bank); shoreline development; septic systems; atmospheric loads; forestry; ocean upwelling; legacy nutrients stored in bottom sediments; and natural background surface runoff and groundwater loadings.

3.1.3 Linking Nutrient Loads to Environmental Response

Understanding the linkages between nutrient loads and environmental response for a specific waterbody relies on an understanding of key attributes of the system. These include

- Spatial scale of the assessment and relevant hydrologic and hydraulic characteristics—the chemical and biological response of a system to nutrient loads depends greatly on the size of the system and the flow regime. Shallow lakes respond differently than deep lakes, small rivers different than large rivers, rivers different than reservoirs, estuaries different than rivers, and so on. Understanding dilution, transport, and residence times of nutrients within a system is required;
- Temporal characteristics—variability from year-to-year, season-to-season, and day-to-day in terms of nutrient loads and water quality response is important. Also, defining critical conditions that are most likely to result in an unacceptable water quality response is needed. Temperature, precipitation, flows, and lake levels will all play an important role. Climate change should also be considered when assessing future conditions;
- Algal growth and loss dynamics—temperature, nutrients, and light are key drivers for algal growth. In some systems, nutrients in addition to nitrogen and phosphorus need consideration, such as carbon and silica. Assessment of multiple algal taxa may be desired. In addition to algae die-off, consideration of scour and zooplankton grazing may be warranted. Invasive species that disrupt the natural ecosystem may also need to be considered;
- Interactions with bottom sediment—bottom sediments can be an important source and/or sink of nutrients. Increased algal productivity leads to additional organic matter and nutrients settling in the sediment layer. Decomposition of the organic matter leads to increased sediment oxygen demand. Low oxygen levels overlying the sediments leads to a release of nutrients back to the water column; and
- Uptake of nutrients by other flora—in some aquatic situations, desirable vegetative growth may consume and store significant amounts

of nutrients, such as large growths of water hyacinths, thus reducing the nutrients available for driving algal growth.

The various factors involved in linking nutrient loads to water quality response can be complex. In all situations, assessments should begin with simple approaches and add complexity only as needed to answer the management questions at hand.

3.2 Modeling to Understand Load-Response Relationships

Models that simulate water quality response to nutrient loads play a valuable role in informing management decisions related to achieving water quality goals and targeting nutrient load reductions. Available models range from strictly empirical statistical relationships to models that combine mathematical formulations of cause–effect linkages. Models are never perfect, but have significant utility despite their limitations because they inform understanding of the site-specific load–response relationship. Water resource managers should strive to understand the model being used, including its limitations, and ensure the necessary data are available to develop and apply the model in a way that is scientifically defensible. Modeling methodology and results will need to be accepted by regulatory agencies, and may face scrutiny by various stakeholder groups.

The selection of the most appropriate model framework for a given site should consider the balance between the rigor of the management questions being asked; the site-specific physical, chemical, and biological characteristics of the system and its complexity; and the resources available to support data collection, model development, and model application. Starting with the simplest approach and building complexity only as needed to inform management decisions is often the best approach.

The Water Environment Research Foundation (WERF) developed practical, model-based approaches and guidance for deriving quantitative relationships between nutrient loads and site-specific water quality and ecological response indicators (WERF, 2013). Other modeling guidance and resources are available from U.S. EPA, the U.S. Army Corps of Engineers, the U.S. Geological Survey, and academic institutions. State regulatory agencies will often have preferred modeling approaches and should be consulted.

3.3 Data Requirements and Assessment

Any effort to understand the water quality response to nutrient loads will be based on the available data for the system. Observed data are needed for both nutrient loads and water quality response. As a first step, existing data should be compiled and analyzed. Analyses may include spatial and temporal plots, assessments of correlation between data, statistical analysis of trends, and comparisons to literature values, among others. Typical types of data to be considered include

- Physical and chemical information—stream flows, temperature, salinity, and so on;
- Nutrient concentrations—typically nitrogen and phosphorus in their various forms and potentially carbon and silica;
- Ambient dissolved oxygen—spatial and temporal variations should be considered;
- Biological data—chlorophyll-*a* is common, but also algal population distributions by taxa and biomass estimates. In some systems, fish and benthic macroinvertebrate community information may be useful;
- Evidence of contribution to algal toxins;
- Robust upstream and downstream data; and
- Data characterizing nutrient loads from the various sources to the waterbody, including watershed land uses and soils.

The existing data set should be evaluated for gaps, and a monitoring plan should be developed and implemented to address the gaps. As with modeling, the desired robustness of the data set should be driven by the management questions being asked, the site-specific characteristics of the system, and the resources available. For a scientifically credible assessment of water quality response to nutrient loads, data sets should be developed applying accepted sampling and analytical methodologies and quality assurance protocols and should be maintained in a reliable data management system.

3.4 Developing Nutrient Load Targets

Once the nutrient load water quality response relationships for a specific system are understood and can be evaluated using a modeling tool or data assessment methodology, they can be applied to refine water quality goals and the nutrient loads required to meet those goals. The data assessments and tools can also be applied to allocate loads among sources based on cost-effectiveness and reliability, and to establish permit conditions for point-source discharges.

Once a target nutrient load is established for a point-source discharger, permit conditions in the form of numeric effluent limitations may need to be developed. The typical approach for development of water-quality-based effluent limitations for toxics based on U.S. EPA guidance does not directly apply to nutrients. Water resource recovery facilities should work with their permitting agency to consider the site-specific load-response relationships to establish permit conditions for the discharge of nutrients. The National Association of Clean Water Agencies presented an assessment of the considerations that should be made when developing effluent limitations for nutrients (Brown and Caldwell, 2014).

4.0 EVALUATE WATER RESOURCE RECOVERY FACILITY EXISTING PERFORMANCE

While state agencies typically take the lead in establishing loading targets, required reductions, and permit limits, WRRFs and utilities should be proactive in their involvement in an effort to ensure that the best available science is applied to inform decision-making related to their planning needs. With the requirement to meet ever lower nitrogen and phosphorus effluent limits, the utility manager needs to understand the nitrogen and phosphorus removal achieved and the optimization and reclamation potential. To do this, facilities are implementing more detailed data collection and analysis, online sensor technology, process modeling, and other means to achieve these limits, while optimizing processes to minimize the use of resources, energy, and capital (Yi et al., 2014). Evaluation of the performance of the existing WRRF and use of operations tools can provide facility personnel and decision-makers with a better understanding of the process changes and opportunities to enhance nutrient removal and recovery and reduction of GHG emissions. This section provides the basis for facilities to improve on saving energy or chemicals or reduce effluent nutrients with the existing infrastructure.

4.1 Robust Data for Influent, Effluent, Sidestreams, and Biosolids

Collection of robust and comprehensive information on monitoring parameters, which are key in decision-making (nitrogen species, phosphorus species, and carbon species), influent, industrial discharges, effluent, unit processes, internal recycle streams, sidestreams, and biosolids, is essential in evaluating WRRF performance assessment. Water Environment Research Foundation (2010) considers frequency of sampling. This set of data is more comprehensive than what is required for compliance.

A robust analysis of the historical data in influent, effluent, unit processes (i.e., biological tanks), recycle streams, sludge management systems, digester supernatant, and centrate, along with the historical facility operational data, is essential, as are mass balances to identify any potential or systematic data collection issues. Integration of real-time data collection (i.e., sensor outputs) from supervisory control and data acquisition systems with operational tools is critical for response to dynamic conditions, process control, facility optimization, and operations planning. Sampling and data analysis should include extreme wet and dry weather conditions and other diurnal, weekly, or seasonal conditions to properly characterize the influent wastewater and dynamic performance. It is recommended to include 24-hour composite sampling for the influent and effluent and grab samples from the biological nutrient removal system (biological tanks) for a nitrogen and phosphorus profile study. Evaluation of WRRF performance cannot be performed without information on the nutrient concentration in the sidestreams and the effect on the raw influent concentrations (Yi et al., 2014).

4.1.1 Nitrogen Species

Influent nitrogen may be present as organic nitrogen (NH_2^-) or inorganic nitrogen species, ammonia, nitrite, and nitrate (NH_3, NO_2^-, and NO_3^-), or other minor intermediate forms. Organic nitrogen associated with complex organics (such as urea) and ammonia generally predominate in the influent. Nitrogen is present in both the liquid and solids portions of the influent. As treatment breaks down complex organics, the NH_2^- molecule is released, increasing the NH_3 concentration in the liquid stream. For many WWRFs, there is an ammonia limit, and, as such, treatment oxidizes ammonia first to nitrite and then nitrate through a biological process termed *nitrification*. Some WRRFs remove nitrogen from the system by conversion of nitrate (or nitrite) to nitrogen gas (i.e., denitrification). The mass of nitrogen is conventionally expressed as the nitrogen component only of each compound. Total nitrogen is the sum of nitrogen concentration or mass in all forms.

Assessment of the removal or recovery of nitrogen through treatment requires an understanding of the wastewater characteristics of all forms of nitrogen, especially influent total Kjeldahl nitrogen (TKN), organic nitrogen, and ammonia. The chemical oxygen demand (COD)-to-nitrogen ratio in the influent is a critical parameter in the operation of a WRRF; therefore, knowing the carbon mass with respect to available nitrogen for cell synthesis is important. The nitrogen content in the sidestreams (e.g., digester supernatant or centrate) is typically high (800 to 1200 mg/L) (Constantine, 2006) and represents essential information to close the mass balance for nitrogen species. Nitrogen data should be developed at key points in the treatment process to fully evaluate existing and potential process removal and recovery and to set the stage for process modification if needed.

The importance of species characterization has increased in the past decade with more stringent effluent nutrient discharge criteria. For nitrogen species, the dissolved organic nitrogen (DON) concentration in the effluent from biological nutrient removal (BNR) treatment facilities ranges from 0.50 to 2.5 mg/L (Pagilla, 2007), which poses a challenge for the WRRFs required to meet low total nitrogen limits of 3 or 5 mg/L. Availability of historical data on influent dissolved organic nitrogen provides useful information to utility owners in meeting the stringent discharge criteria. Measurement of nitrate and nitrite in the anoxic tank provides insight on the degree of denitrification and sufficiency of the carbon source for complete denitrification. Historical data on alkalinity in the influent and effluent allows for mass balance calculation for alkalinity consumption and recovery through nitrification and denitrification, respectively. Anaerobic digested sludge creates a reject water stream that can contribute 15 to 20% of the ammonia load recycled to the liquid train of a WWRF (Constantine, 2006).

4.1.2 Phosphorus Species

Influent and effluent total phosphorus and soluble phosphorus measurements are critical in evaluation of a WRRF in terms of phosphorus removal. Phosphorus may be removed or recovered by incorporation to biomass, through chemical precipitation and physical removal, or both. Water Environment Research Foundation (2010) recommends sampling of total phosphorus and ortho-phosphorus (ortho-P) once a week; however, availability of historical data is important for planning. Soluble phosphorus (ortho-P) measurements in anoxic and aerobic tanks provide data for calculation of phosphorus uptake and recovery in enhanced biological phosphorus removal (EBPR) facilities. Analysis of available carbon (in forms of volatile fatty acids [VFAs]) is also important in evaluating EBPR at the facility and opportunities for optimization and/or future planning.

Water resource recovery facilities are able to recover phosphorus for commercial fertilizer production from wastewater. Currently, some WRRFs are having success with struvite recovery, which allows for the precipitation and recovery of both nitrogen and phosphorus. Characterization of the sidestream of sludge treatment and sludge liquor, where the phosphorus concentration is high (such as struvite recovery) is important. Data on the chemical dosage for phosphorus removal also allow for better understanding of phosphorus removal through current secondary treatment processes.

Soluble nonreactive phosphorus (SNRP) is similar to refractory organic nitrogen in terms of the type of compounds, significance, and environmental fate and effect. The concentration of SNRP in the influent is also important for WRRFs required to meet extremely low total phosphorus concentrations. Based on a body of research to date, the concentration of SNRP is case-specific to each WRRF and, therefore, available data on the total phosphorus fractionation in the influent provide useful insight for future planning to achieve lower total phosphorus limits in the effluent.

4.1.3 Carbon

For BNR, it is important that the activated sludge has enough available carbon for denitrification. Readily biodegradable COD is required to complete denitrification, which is available in the raw influent COD or external carbon sources. Biological phosphorus (bio-P) removal also strongly depends on the availability of VFAs in the raw influent, which serves as a carbon source for the bacteria responsible for bio-P removal. Sufficient historical data on COD, biochemical oxygen demand (BOD), and soluble COD and BOD in the influent allow for optimization of the nutrient removal processes in terms of COD:TKN and COD:total phosphorus. Analysis of the available carbon in the influent for nutrient removal, along with the target nutrient removal, enable the WRRF to make decisions on the quantity and type of supplemental carbon source required for nutrient removal.

4.1.4 Flows

Understanding influent flow and dynamics is important to the operation of the WRRF and for planning purposes. Based on historical data on the flows, annual average daily flow, maximum monthly peak flow, and peak daily flow can be obtained. The hydraulic loading factors on diurnal 7-, 15-, and 30-day rolling averages for use of probability distribution functions can also be obtained. If possible, WRRFs must adjust to the diurnal variations in flow. As ammonia and soluble phosphorus concentration are returned to the treatment facility through the liquid stream from the sludge management system, such as gravity thickener overflow, centrate, and digester supernatant and internal recirculation, it is important to obtain information on the flows for evaluation of WRRF flow and mass loading. Flow measurement for the waste activated sludge, return activated sludge, and nitrified mixed liquor return streams should be considered for sludge age calculation and process optimization.

Accurate process modeling studies strongly depend on the availability of sufficiently long influent time series (representing the inherent natural variability at the facility inlet as accurately as possible). Additional sampling and laboratory analysis to characterize the influent wastewater is needed to fulfill model input requirements in terms of dry and wet weather flows, diurnal flows, seasonal and weekly variations in the flow, and septage/industrial wastewater loads. The fractions of each of the wastewater components should be understood for various conditions. Utilization of facility-wide models that are calibrated with sufficient historical data provides a useful tool for better understanding the facility processes for process control, optimization, and planning and are further discussed in Section 6.0 of this chapter.

4.1.5 Identify and Fill Data Gaps

Analysis of the inventory data collected for flow, nutrients, and carbon in influent, effluent, sludge management processes, and other streams will allow for recommendations to be made to identify the data gaps. It is important to identify critical data gaps related to influent characteristics, unit process performance, and additional data required for model development and calibration procedures. It is important to target collection of the needed data on influent characterization (nitrogen, phosphorus, and carbon species), chemical dosage, sludge retention time, sensor data, sidestream flows, and operational parameters such as temperature and dissolved oxygen.

4.2 Evaluate Averages, Ranges of Values, Daily, Weekly, and Seasonal Loads

Assessment of the daily loads, average daily loads, daily design load, peak hourly load, maximum daily load, maximum monthly load, peak instantaneous load, and load equalization is important in evaluating hydraulic and nutrient loading of the WRRF to develop process control strategies and unit process sizing for improved operation, optimization, and future planning.

4.3 Determine Removal/Recovery

Water resource recovery facility treatment performance is characterized by degree and extent of nutrient removal/recovery. Information on the WRRF influent and effluent nutrient content and key unit processes' influent and effluent nutrient content is required to determine the removal/recovery rate for the overall WRRF and each key unit process. Determination of the removal/recovery of nutrients at the WRRF is the first step in defining operational/process changes required for higher degrees of nutrient removal/recovery.

4.3.1 Statistical Analysis

Monitoring information based on average load provides the best readily available information with which to characterize WRRF performance in terms of nutrient recovery/removal. Selection of a data-averaging period is important. The average of reported daily/monthly/annual average nutrient loadings indicates long-term treatment performance. The monthly average values may not be representative of daily fluctuations in nutrient removal/

recovery by WRRFs, and the actual daily data may be required to better characterize the performance. Reliability data graphs for nutrient removal can be developed for weekly, monthly, and annual averages (weekly, monthly, and annual frequency curves). Median/mean, standard deviation, and coefficient of variation (COV) are also useful tools for process evaluation (U.S. EPA, 2007).

4.3.2 Variability

Some variations will be observed in the performance data and nutrient removal/recovery because of process upsets caused by weather conditions, accidents, and equipment failure. Other variations will be attributable to natural variations in influent conditions and natural variability in laboratory measurements, sampling, and flow. The range of reported data based on daily, seasonal, and annual load is also important to indicate long-term performance of a WRRF in terms of nutrient removal/recovery. Reliability data graphs for nutrient removal can be developed for the actual data for weekly, monthly, and annual sampling to obtain frequency curves. Median/mean, standard deviation, and COV can also be obtained (U.S. EPA, 2007).

4.4 Energy Considerations

Nutrient removal processes may cause increased energy use and change energy recovery at the WRRF. Energy effects are case-specific and strongly depend on the processes involved at the WRRF. Assessment of energy use at a WRRF allows consideration of energy effects with respect to increased nutrient recovery or removal. Developing a strategy to regularly monitor energy use in the WRRF unit processes will allow for continuous improvement in terms of energy use inventory and the potential for energy recovery.

4.4.1 Usage by Unit Process

Based on available data on energy use by unit process for electric load profiles, equipment performance curves, and normal operating time for intermittently operated processes (Santos et al., 2008) valuable information can be obtained on the existing performance of each unit process in terms of energy use. The energy use can be obtained based on unit energy per unit of BOD removed. Evaluation of energy use over time at a WRRF includes identification of fluctuations and variations in energy use and monitoring the effect of energy-efficiency improvements.

4.4.2 Energy Production

Energy neutrality is a goal for many WRRFs. The energy content of wastewater is in four dominant forms: chemical, thermal, kinetic, and potential energy in the form of liquids and solids. The main potential for energy recovery from WRRFs in treatment processes via anaerobic digestion with biogas utilization and biosolids incineration with electricity generation or other energy recovery strategies could help offset the electricity consumption of the WWRF (Stillwell et al., 2010). The rate of energy production should be examined based on the unit of energy per mass of biosolids/solids used. The extent of energy recovery and potential for sustainability is a function of the scale of the WRRF. Analysis of energy balance is important in the evaluation of nutrient recovery or removal processes in terms of energy production.

4.5 Estimate Greenhouse Gas Emissions

Long-term sustainability is also a goal for many utilities, which includes recognition of and reduction in GHG emissions. Protocols for estimating GHG emissions from municipal wastewater sources have been developed by WERF. Predominant direct sources of WRRF-related GHG emissions are the result of purchased electricity, stationary combustion (including flares, boilers, and cogeneration systems), treatment processes, and fleet vehicle fuel combustion. Several organizations have established voluntary GHG emissions with standards that have set the basis for regulatory emissions reporting protocols, such as the World Resources Institute/World Business Council for Sustainable Development (2005) GHG Protocol, The Climate Registry (TCR), Local Government Operations Protocol (TCR, 2010), and the International Organization for Standardization (2006) 14064 GHG standard. Water Environment Research Foundation (2009) and other researchers are gathering field data to develop improved estimating methods and emissions factors that can be used to more accurately characterize local facility-level emissions based on specific operating parameters (e.g., Chandran et al., 2009).

All of these protocols follow the same general approach. Basic steps for preparing organization-level GHG emissions inventories include selecting a base year, establishing inventory boundaries, identifying and categorizing facility GHG emissions sources, and completing the GHG inventory and quantifying emissions. Emissions calculations are based on available activity data, emissions factors (published, calculated, or measurement-based), and application of global warming potentials.

5.0 SOURCES AND CHARACTERISTICS OF NUTRIENTS TO THE WATER RESOURCE RECOVERY FACILITY

Sources of nitrogen and phosphorus to the WRRF should be determined to understand the variability in nutrient loading and sources that may be amenable to source control. Typical domestic wastewater will contribute 12 g total nitrogen and 3.6 g total phosphorus per capita per day (www.ces.ncsu.edu/plymouth/septic3/MinnisNutrientsText.pdf). If the population changes during the year (as in university or seasonal resort communities), those fluctuations in the contributing population should be understood. Communities that have combined sewers may receive some additional nutrient loading from surface runoff captured and conveyed for treatment.

Industrial wastewater may contribute significant nitrogen and phosphorus that may not have been subject to pretreatment standards or industrial surcharges. For instance, process wastewater from food processing operations may include product residuals and nitric or phosphoric acid equipment cleaners. Nitric acid is commonly used as a cleaner in metal industries cleaners; the WRRF may receive a neutralized process wastewater subject to categorical or local limits for metals, but not for the nitrate. The source of hauled waste ranges from septage similar to domestic wastewater to high-strength industrial wastewater. The wastewater utility generally has discretion with respect to acceptance of hauled waste. Hauled waste may be processed through the entire liquid treatment train or introduced to anaerobic digestion and contribute to return sidestreams. In any case, the utility needs to fully characterize accepted outside waste including nutrient loading.

Wastewater characteristics may change in the collection system. Long travel times from the source to the WRRF may transform complex compounds with organic nitrogen to ammonia. Anoxic conditions in the collection system may result in some denitrification during conveyance. Water conservation programs and programs to reduce clear water intrusion to the system may change the concentration (not the loading) from sources. Increased nutrient concentration may be a consideration in recovery process controls.

6.0 DEVELOP PROCESS MODEL TO EVALUATE PROCESS MODIFICATIONS/ADDITIONS TO MEET PROPOSED OR ALTERNATE EFFLUENT LIMITS/PERMIT CONDITIONS

A key step in evaluating process modifications/additions to meet proposed effluent limits is leveraging available mathematical computer facility simulators. There are several commercially available software modeling packages that simulate optimization and process additions, such as GPS-X® by Hydromantis, BioWin® by Envirosim, and others.

The following subsections discuss a systematic approach for reducing nutrient loads at a WRRF and the role of the process model for such efforts.

6.1 Evaluate, Select, and Utilize a Process Model

There should be a systematic approach for using a commercially available modeling software package. The approach should be based on the adage, "beginning with the end in mind". Refer to the Water Environment Federation Manual of Practice No. 31 titled *Wastewater Treatment Process Modeling* (2014) and Rieger et al. (2012), which have a framework for good modeling practice. The frameworks have a prescribed methodology for producing goal-oriented results. In both publications, there are several case studies that illustrate the required facility understanding, data quality, and quantity, and the effort for calibration/validation with respect to a defined goal. Once the project goals and approach are finalized, a process model can be custom-built and calibrated against historical WRRF performance data.

Once constructed, the customized simulator can be leveraged to evaluate a suite of optimization and process addition scenarios. Evaluating nutrient reduction strategies at WRRFs should balance the practical with the visionary. The practical refers to proven, tried, and reliable nutrient reduction strategies/technologies; and visionary refers to emerging and embryonic technologies, innovative strategies, and so on.

6.2 Optimization of Existing Systems

Optimization of existing facilities is the first step toward nutrient reduction. Nutrient removal by optimization might be possible at processes within the WRRF that operate below design load. Any unused process capacity might be devoted to nutrient reduction until loads reach capacity levels. Nutrient removal takes advantage of unused tankage, new process approaches, instrumentation improvements, and, without a permit limit with potential enforcement penalties, gets as much nutrient reductions as possible.

The strategies for optimization strive to be simple, low-capital-cost improvements that can be implemented quickly. The strategies can be grouped into "no-capital-cost" and "low-capital-cost" strategies; a sample of potential strategies is listed in Table 3.1.

The calibrated process model can be used to confirm/validate any of the proposed facility optimization concepts listed in Figure 3.1.

6.3 Sidestream Treatment and Nutrient Recovery

The *sidestream* refers to the return streams from biosolids processing. Despite their small flows (typically less than 5% of raw plant flow), the

TABLE 3.1 Nutrient optimization examples.

No-capital-cost strategies	Low-capital-cost strategies
Use offline tankage to provide additional treatment	Add instruments for nutrient removal in ammonia-based aeration control mode
Modify operational mode, such as raising the solids residence time	Add chemicals for phosphorus removal
Modify blower operating set points	Add chemicals to reduce load, unlock capacity
Operate in split treatment mode	Add anoxic and/or anaerobic zones for biological nutrient removal
Change to simultaneous nitrification/denitrification operation	Add internal recycle for denitrification
Shut down aeration to create unaerated zones	Add mixers for unaerated zones

sidestream represents about 15 to 40% of the discharge nutrient load, as shown in Figure 3.2.

The benefits of removing nutrients in the sidestream are as follows:

- Warm water and concentrated nutrients (favorable kinetics, small footprint);
- Low flows (ability to equalize);
- Typically more cost-effective in terms of (U.S.) dollars/kilogram nutrient removed than liquid stream treatment;
- Potential for reduced aeration/chemicals than liquid stream treatment (limited to nitrogen removal);
- Provides an opportunity to recover sidestream nutrient loads;
- Easier to phase construction than liquid stream treatment; and
- Might save a WRRF from requiring chemical addition for liquid stream nutrient removal.

Despite all these benefits, not all plants are candidates for sidestream treatment; as such, interested WRRFs need to confirm they are candidates. Facilities with anaerobic digestion, dewatering, and a dedicated dewatering return stream are candidates. Operators of facilities with other configurations should confirm they are candidates before pursuing sidestream treatment.

6.4 Process Addition

The first step in determining process additions necessary to meet different nutrient removal objectives is to understand the WRRF's existing infrastructure and space constraints. Existing infrastructure should be included in future upgrades as much as possible, especially if facilities are less than 10 years old. In cases of severely constrained sites, removal and replacement of existing facilities may be required.

A calibrated WRRF process model can confirm whether the process addition will meet the desired treatment objectives. Besides model results, the user should confirm the findings against previously published data on unit process performance (e.g., Bott and Parker, 2011). Additionally, the model results can highlight any significant facility-wide ripple effects. For example, if a WRRF transitions from secondary treatment to nutrient removal, the process model should capture the effect on biosolids treatment (e.g., yield).

6.5 Effluent Nutrient Loading and Concentration

Most WRRF NPDES effluent permits are both concentration and load based. The issue with using both is that a WRRF gets no credit for diverting loads before discharge. For example, a WRRF that recycles a portion of their flows prior to discharge does not get any nutrient load reduction credit if discharge limits include concentration. Using loads is beneficial because they are independent of the effect on flows (e.g., water conservation) while providing nutrient removal credit for facilities that divert flows before discharge (e.g., recycled water).

6.6 Resource Recovery

Nutrient recovery at WRRFs is gaining traction and is a reality at several facilities. The first step in nutrient recovery is capturing them in biosolids, such as enhanced biological phosphorus removal (EBPR) or cell assimilation in liquid stream treatment. Once captured with biosolids, the nutrient endpoint is dependent on biosolids treatment. For facilities with digestion, some captured nutrients are released with any volatile solids destruction. Such released nutrients are returned to the liquid stream with the dewatering return sidestream.

Typically, nutrients can be recovered in the dewatering return sidestream because of the sidestream attributes discussed previously (e.g., high strength, low flow, etc.). Phosphorus and ammonia are the most common nutrient recovered through chemical precipitation using metal salts or struvite precipitation. Extracted struvite can reduce sidestream orthophosphate loads from 80 to 90% and ammonia from 15 to 30% (Khunjar and Fisher, 2014).

There are several commercially available technologies for nutrient recovery at WRRFs with the emphasis on phosphorus and ammonia recovery via chemical precipitation. Most of the phosphorus recovery installations are for larger facilities (> 150 ML/d, or 40 mgd) because of economies of scale. However, the noneconomic benefits of nutrient recovery should lead to installations at smaller facilities. Nitrogen recovery technologies exist, but they have gotten lesser attention than phosphorus. Depending on the software modeling package, process model simulators can either calculate nutrient recovery and/or force the simulator to recover nutrients.

6.7 Energy and Chemical Considerations

It is well documented that transitioning from secondary treatment to nutrient removal will require additional energy (e.g., Falk et al., 2013). The extent of increase is dependent on the discharge nutrient limits and averaging periods (e.g., average monthly versus maximum daily). The additional demands for nutrient removal become more pronounced once an external carbon source and filters are required to meet the more stringent limits.

Depending on the software modeling package, some process model simulators quantify energy and chemical demands. In most cases, it is easier to separately calculate the effect on energy and chemical demands as this is site-specific.

6.8 Greenhouse Gas Emissions

An overview of the approaches for calculating greenhouse gas (GHG) emissions was provided in Section 4.5. The initial GHG emissions studies at WRRFs focused primarily on energy and chemical consumption (e.g., Pagilla et al., 2009). Such studies found that the majority of GHG emissions at WRRFs are associated with energy demand (approximately 55 to 65%) (Falk et al., 2013). This study considered the effect on GHG emissions for various effluent nutrient levels, as shown in Figure 3.2. The results suggest that GHG emissions increase with more stringent nutrient limits as a result of increased energy and chemicals.

Future GHG emissions studies at WRRFs will more accurately quantify the role of nitrous oxide and methane emissions to assist in the decision-making process. The commercially available process simulator software packages such as BioWin® 4.0 are starting to include GHG emissions calculators from carbon dioxide, methane, and nitrous oxide emissions. Despite uncertainties in emissions levels, the field is continuing to advance.

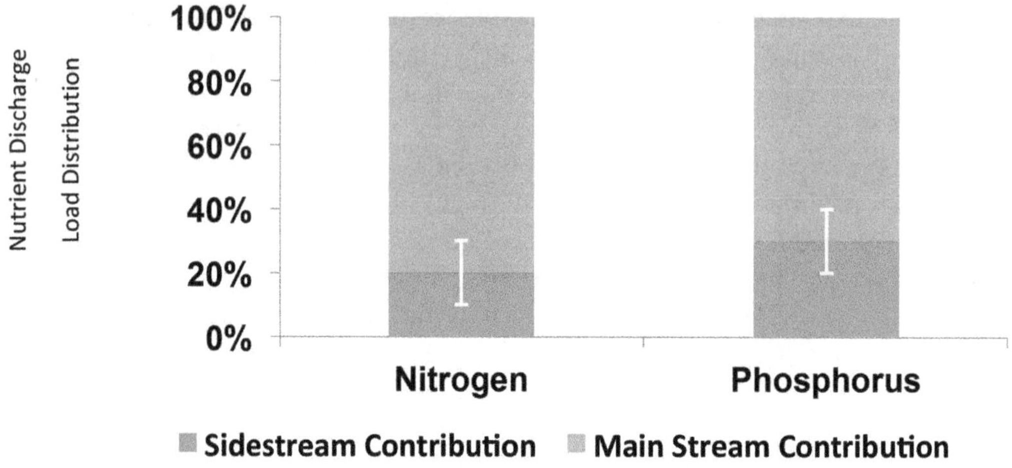

FIGURE 3.2 Typical nutrient discharge load contributions.

6.9 Cost

Most of the commercially available software packages do not calculate costs. Several other software packages can be used in tandem with a software model, such as CapDet Works®. However, costs associated with facility upgrades for WRRFs are extremely site-specific and, as such, caution must be taken for such an approach. Figure 3.3 illustrates the general trend for greater unit removal cost for nitrogen with increasing treatment processes.

7.0 ALTERNATIVES AND DECISION-MAKING

This chapter identifies a multitude of factors that should be weighed by the utility manager when determining the long-term plan for nutrient recovery or removal. The regulatory requirements are only the first step in the evaluation. The utility will need measurable tools to make informed and publically acceptable decisions. The goals and objectives for recovery/removal should be clearly established at the beginning of the evaluation process. These tools can be as simple as comparative matrices to sophisticated production theory curves. For most projects, there will be multiple objectives, and the importance of each must be understood and agreed by stakeholders at the outset of the planning process. To help the reader address all of the important factors, the key considerations are presented as a series of questions in the following subsections.

7.1 Process Modifications

Is there a recovery/removal process that will be compatible with existing processes? Is there room for the new process? What are the capital and

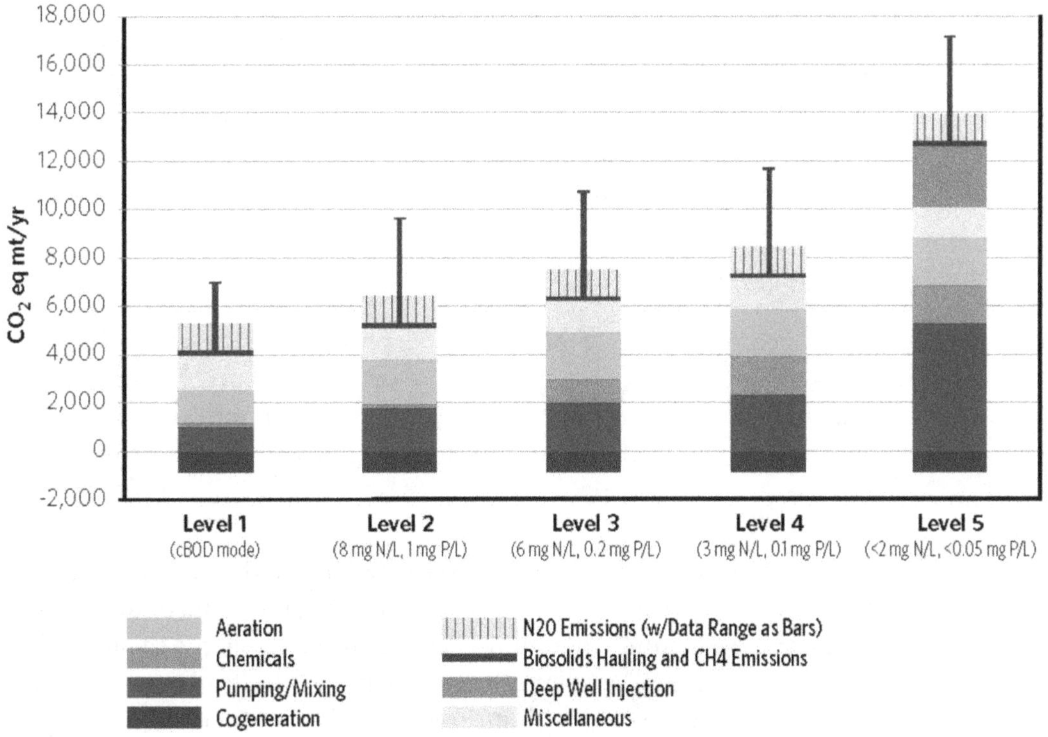

FIGURE 3.3 Greenhouse gas emissions distribution for various nutrient discharge limits (adapted from Falk et al. [2013]).

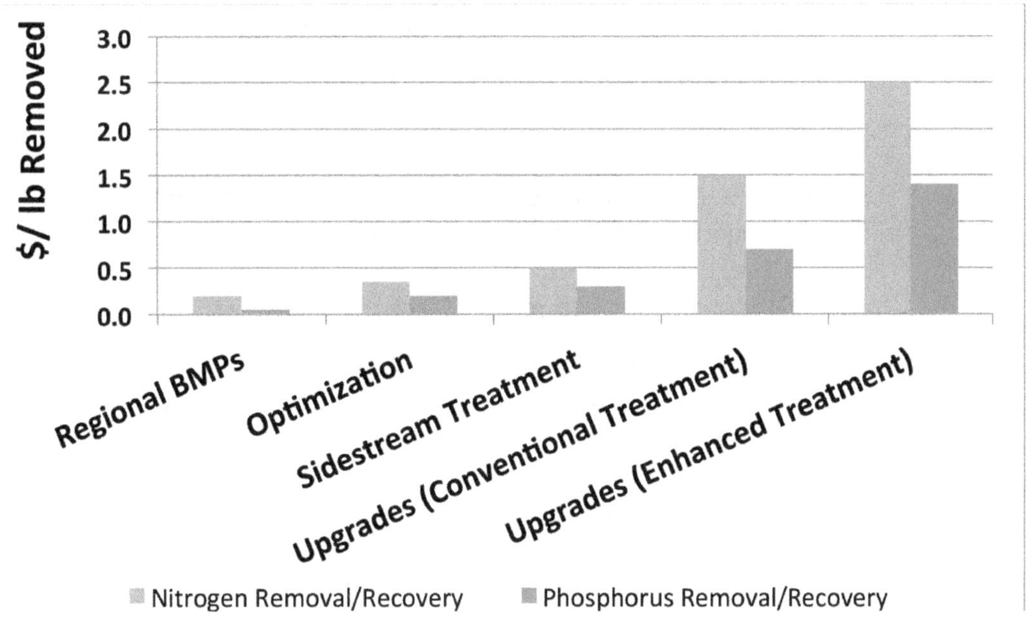

FIGURE 3.4 Hypothetical unit cost comparison for various nutrient load reduction strategies.

operational costs? Is there cost recovery? Are the process modifications flexible for future regulatory changes? Are qualified operators available to manage the additional processes? If more phosphorus is removed to biosolids, will application at a phosphorus limit rather than a nitrogen limit make that program less attractive to agricultural partners? How do the process modification options coincide with other goals such as energy neutrality, cost-of-service objectives, and sustainability?

7.2 Source Control

Can significant sources of nitrogen and phosphorus be controlled at a source? Is it more cost-effectively removed at the source? What are the revenue implications for the utility if there is source removal? If the source removes nutrients, what are the additional watershed implications?

7.3 Trading

Is trading an allowable option in the state? What are the allowable trading partners (e.g., watershed hydrologic unit, instate, and multistate)? Longevity of the trade? Trading conditions? Who is responsible for the nonutility reductions? Is there an exchange mechanism for the trade? Is the utility located within a watershed with upstream trading partners?

7.4 Adaptive Management

Is there an adaptive management option within the state regulations? Are there other stakeholder contributors that will collaborate for overall nutrient concentration reductions in the affected watershed? How will the success of adaptive management be evaluated?

7.5 Seasonal Limits

Does the effect of nutrient discharge vary by season (wet and dry seasons and winter–summer differences)?

7.6 Variance (Socioeconomic Effects)

Will the utility modifications and long-term operational costs be affordable to ratepayers? Is an individual or area-wide variance institutionally available? What are the criteria?

7.7 Loadings Versus Concentration Permit Conditions

Are there local receiving stream water quality criteria that must be met to correct an identified impairment of the designated aquatic life or drinking water use? Or, is the utility reducing nutrient discharges to meet a larger regional (downstream reservoir) or overall watershed (Mississippi River Gulf hypoxia area) reduction goal?

7.8 Informed Decisions

With the aforementioned information and evaluations, the utility should then be in a position to select a pathway for nutrient recovery or removal that optimizes the following:

- Removal/recovery/reuse,
- Watershed effects,
- Resource recovery,
- Energy use/production,
- Greenhouse gas minimization,
- Cost, and
- Other.

The final nitrogen and phosphorus effluent limits and other related permit conditions for the utility should only be accepted after the utility has a clear understanding of its ability to meet those conditions and the other factors listed previously.

8.0 REFERENCES

Bott, C. B.; Parker, D. S. (2011) Nutrient Management Volume II: Removal Technology Performance & Reliability. Water Environment Research Foundation NUTR1R06k; IWA Publishing: London.

Brown and Caldwell (2014) *Review of U.S. EPA Methods for Setting Water Quality-Based Effluent Limits for Nutrients*; Prepared for the National Association of Clean Water Agencies, Washington, D.C.

Camargo, J. A.; Alonso, A.; Salamanca, A. (2005) Nitrate Toxicity to Aquatic Animals: A Review with New Data for Freshwater Invertabrates. *Chemoshere*, **58**, 1255–1267.

Chandran, K.; Ahn, J. H.; Kim, S.; Park, H.; Katehis, D.; Pagilla, K. (2009) Spatial and Temporal Variability in N_2O Generation and Emission from Wastewater Treatment Facilities. *Proceedings of the Water Environment Federation Specialty Conference on Nutrient Removal: Sustainable Treatment Solutions* [CD-ROM]; Washington, D.C., June 28–July 1; Water Environment Federation: Alexandria, Virginia.

Chesapeake Bay Commission (2012) *Nutrient Trading for the Chesapeake Bay. An Economic Study*; Chesapeake Bay Commission: Annapolis, Maryland.

Constantine, T. A. (2006) North American Experience with Centrate Treatment Technologies for Ammonia and Nitrogen Removal. *Proceedings of the Water Environment Federation 79th Annual Water Environment Federation Technical Exhibition and Conference* [CD-ROM]; Dallas, Texas, Oct 21–25; Water Environment Federation: Alexandria, Virginia.

Falk, M. W.; Reardon, D. J.; Neethling, J. B.; Clark, D. L.; Pramanik, A. (2013) Striking the Balance between Nutrient Removal, Greenhouse Gas Emissions, Receiving Water Quality, and Costs. *Water Environ. Res.,* **85** (12), 2307–2316.

Hickey, C. W.; Martin, M. L. (2009) *A Review of Nitrate Toxicity to Freshwater Aquatic Species*; Report No. R09/57; Environment Canterbury: Timaru, New Zealand.

International Organization for Standardization (ISO) (2006) Greenhouse Gases—Part 1: Specification with Guidance at the Organization Level for Quantification and Reporting of Greenhouse Gas Emissions and Removals; ISO 14064-1; International Organization for Standardization: Geneva, Switzerland.

Khunjar, K.; Fisher, J. (2014) Nutrient Recovery in the Global Water Industry. From the Water Environment Association of Texas (WEAT) 2014 Eckenfelder Lecture Series. Hazen and Sawyer; CH2MHill; and Water Environment Research Foundation.

Minnis, P. (Undated) Sources of Nutrients in Wastewater. www.ces.ncsu.edu/plymouth/septic3/MinnisNutrientsText.pdf (accessed Feb 2015).

Pagilla, K. (2007) Organic Nitrogen in Wastewater Treatment Plant Effluents. Presentation at the *STAC/WERF Workshop: Establishing a Research Agenda for Assessing the Bioavailability of Wastewater Derived Organic Nitrogen in Treatment Systems and Receiving Waters*; Baltimore, Maryland, Sept 28.; Water Environment Research Foundation: Alexandria, Virginia.

Pagilla, K.; Shaw, A.; Kunetz, T.; Schiltz, M. (2009) A Systematic Approach to Establishing Carbon Footprints for Wastewater Treatment Plants. *Proceedings of the 82nd Annual Water Environmental Federation Technical Exposition and Conference* [CD-ROM]; Orlando, Florida, Oct 10–14; Water Environment Federation: Alexandria, Virginia.

Rieger, L.; Gillot, S.; Langergraber, G.; Ohtsuki, T.; Shaw, A.; Takacs, I.; Winkler, S. (2012) *Guidelines for Using Activated Sludge Models*; International Water Association Task Group on Good Modelling Practice: London, U.K.

Santos, A.; Lampman, G.; O'Connor, K. (2008) *New York State Energy Research and Development Authority (NYSERDA) and Strategic Energy Management at Municipal Wastewater Treatment Facilities, Clear Waters*; pp 15–16.

Stillwell, A. S.; Hoppock, D. C.; Webber, M. E. (2010) Energy Recovery from Wastewater Treatment Plants in the United States: A Case Study of the Energy-Water Nexus. *Sustainability,* **2** (4), 945–962.

The Climate Registry (2010) *Local Government Operations Protocol, Version 1.1*; The Climate Registry: Los Angeles, California.

U.S. Environmental Protection Agency (2007) *Advanced Wastewater Treatment to Achieve Low Concentration of Phosphorus*; U.S. Environmental Protection Agency: Washington, D.C.

U.S. Environmental Protection Agency (2010) *Final Report on Acute and Chronic Toxicity of Nitrate, Nitrite, Boron, Manganese, Fluoride, Chloride, and Sulfate to Several Aquatic Animal Species*; EPA-905/R-10-002; U.S. Environmental Protection Agency: Washington, D.C.

U.S. Environmental Protection Agency (2013) *Aquatic Life Ambient Water Quality Criteria for Ammonia – Freshwater;* EPA-822/R-13-001; U.S. Environmental Protection Agency: Washington, D.C.

Water Environment Federation (2010) *Technical Submission to EPA: Removal of Nutrients with Currently Available Secondary Treatment Technologies*; Water Environment Federation: Alexandria, Virginia.

Water Environment Federation (2014) *Wastewater Treatment Process Modeling*, 2nd ed.; Manual of Practice No. 31; Water Environment Federation: Alexandria, Virginia.

Water Environment Research Foundation (2009) *Protocols for Estimating Greenhouse Gas Emissions from Municipal Wastewater Sources—Technical Practice Update*; Water Environment Research Foundation: Alexandria, Virginia.

Water Environment Research Foundation (2010) *WERF Nutrients Research Challenge, Nutrient Management Compendium, Operation and Control*; Water Environment Research Foundation: Alexandria, Virginia.

Water Environment Research Foundation (2013) *Modeling Guidance for Developing Site-Specific Nutrient Goals;* Water Environment Research Foundation: Alexandria, Virginia.

World Resources Institute (2013) *Can Nutrient Trading Shrink the Gulf of Mexico's Dead Zone?* World Resources Institute: Washington, D.C.

World Resources Institute; World Business Council for Sustainable Development (2005) The Greenhouse Gas Protocol. http://www.ghgprotocol.org/about-ghgp (accessed Feb 2015).

Yi, P.; Gellner, W. J.; Bilyk, K.; Katimer, R.; Pitt, P.; Kutcher, T.; Riess, R.; Saylor, E. (2014) Operations Tools for the Facility of the Future. *Proceedings of the 87th Annual Water Environment Federation Technical Exhibition and Conference* [CD-ROM]; New Orleans, Louisiana, Sept 27–Oct 1; Water Environment Federation: Alexandria, Virginia.

4

Resource Recovery

1.0 INTRODUCTION

1.1 The Water–Nutrient–Energy Nexus

As a utility moves forward with planning for nutrient reduction in the watershed, it is important to understand the environment, not only with respect to nutrients and meeting water quality standards, as presented in the previous chapter, but also to frame the environment holistically in terms of the nutrient cycle, water use and water reclamation, and energy use and production. As presented in Figure 4.1, with the global population projected to reach 9 billion by 2048, the wastewater treatment industry needs to be cognizant of this interrelationship and consider how it will sustainably feed, provide potable water and electricity, and manage the waste produced by this exponential growth in population.

Utilities need to understand and consider the full spectrum of issues related to the water–nutrient–energy nexus in the wastewater treatment field, including the following:

- The significant amount of power and resources required to remove nutrients from the waste stream;

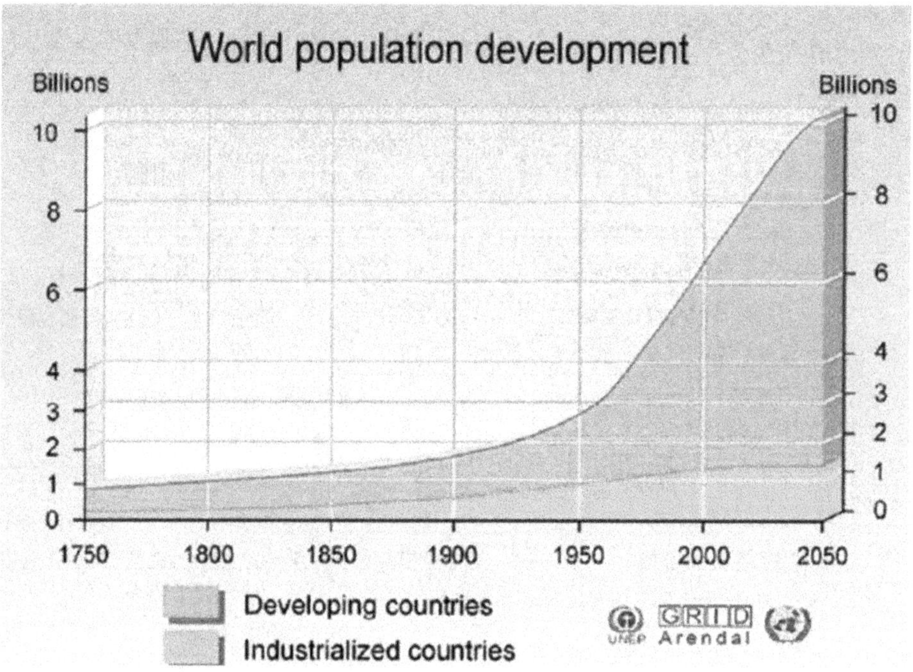

FIGURE 4.1 Historic and projected population of the earth (http://www.sustainablescale.org/areasofconcern/population/populationandscale/quickfacts.aspx).

- The additional power and resources necessary to extract the nutrients removed from the waste stream to generate a reusable product (compared to the manufacturing of other fertilizer products);
- The potential for heat and power generation at a reclamation facility through anaerobic digestion, understanding that this process creates a highly concentrated sidestream with respect to nutrients, which then must be treated or extracted;
- The potential use of reclaimed water in power production (in lieu of potable water) because production of electric power is one of the largest uses of water in the United States;
- The potential use of reclaimed water for irrigation, in agriculture, and in other industrial applications;
- Iron or aluminum salts are required to meet stringent phosphorus limits; however, the use of these chemicals increases the production of residuals and generates inert solids with no British thermal unit (Btu) value that must be managed;
- A supplemental carbon source (in the form of methanol, acetic acid, glycerol, and others) is often required to meet stringent nutrient limits. However, byproducts from the food and beverage or other industries in the service area could be considered as a supplemental carbon source. Sludge fermentation within the facility can also better harness the carbon available in the system; and
- The acceptance of fats, oils, and greases (FOGs) and/or organics at a water resource recovery facility (WRRF) increases the energy production in the digesters and provides a revenue stream for the utility; in the case of organics, however, it also increases the nutrient load at the facility.

All of these factors, and others, need to be balanced to provide the most reasonable, holistic solution to meet water quality standards without resulting in direct or indirect environmental effects that are more severe than the water quality improvements realized. The primary mission of a facility was once solely to improve water quality in the receiving waters. A secondary mission has emerged to meet the water quality standards in a sustainable fashion through energy generation and conservation and reclamation of water and nutrients. The utility must also effectively work with regulators to ensure the National Pollutant Discharge Elimination System (NPDES) limits proposed consider the environment holistically, are reasonable, and do not do more environmental harm than good. Finally, utilities should embrace the concept that they are not only a wastewater treatment facility, but also WRRF, as shown conceptually in Figure 4.2.

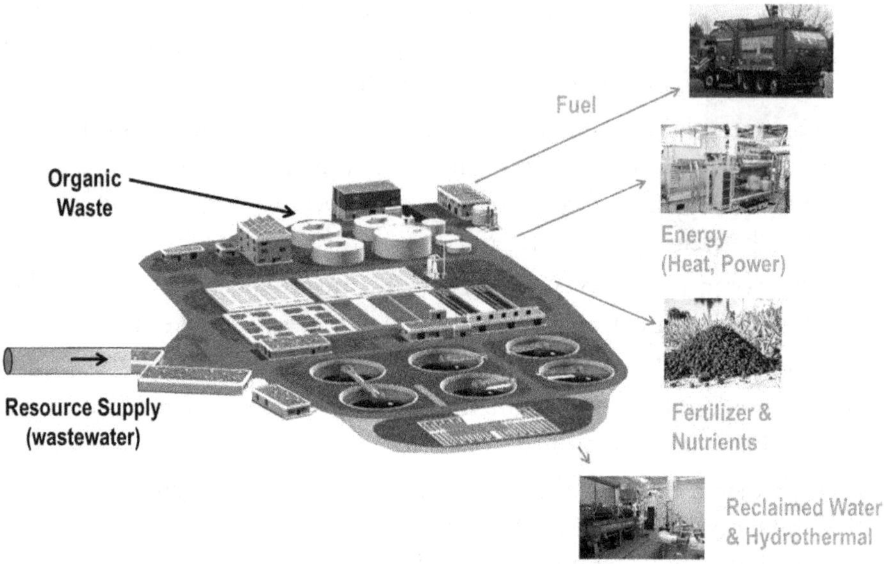

FIGURE 4.2 Wastewater plants are being viewed as resource centers.

1.2 Impetus Behind Nutrient Recovery

Like water and oil, phosphorus is a finite resource on the planet and one that is essential to the agricultural industry and, thus, the production of food for the planet's population. As the population has grown, so has the consumption of phosphorus, as shown in Figure 4.3. Yet 90% of the phosphorus rock reserves is controlled by just five countries (i.e., Morocco and Western Sahara, Iraq, China, Algeria, and Syria). United States reserves constitute about 2% of the world's reserves. Typically, approximately 8 to 10% of U.S. consumption is imported primarily from Morocco (Jasinski, 2012). Despite the recently discovered reserve in Morocco, some have predicted that the phosphorus rock reserves may be depleted in 50 to 100 years (Massachusetts Institute of Technology, http://web.mit.edu/12.000/www/m2016/finalwebsite/problems/phosphorus.html). Therefore, the recovery of phosphorus from the waste stream supplements a critical global need that will become more and more relevant as phosphorus rock demand exceeds the supply.

The implementation of extremely stringent phosphorus limits has inspired phosphorus recovery as an integral part of an overall phosphorus management strategy. In addition, the implementation of anaerobic digestion to stabilize solids, reduce the mass of solids to be handled, and produce biogas for energy recovery has resulted in a high-strength sidestream that actually makes nutrient recovery more cost-effective. As the pieces fall together and a facility moves toward sustainable operations, nutrient recovery becomes one more piece of the puzzle.

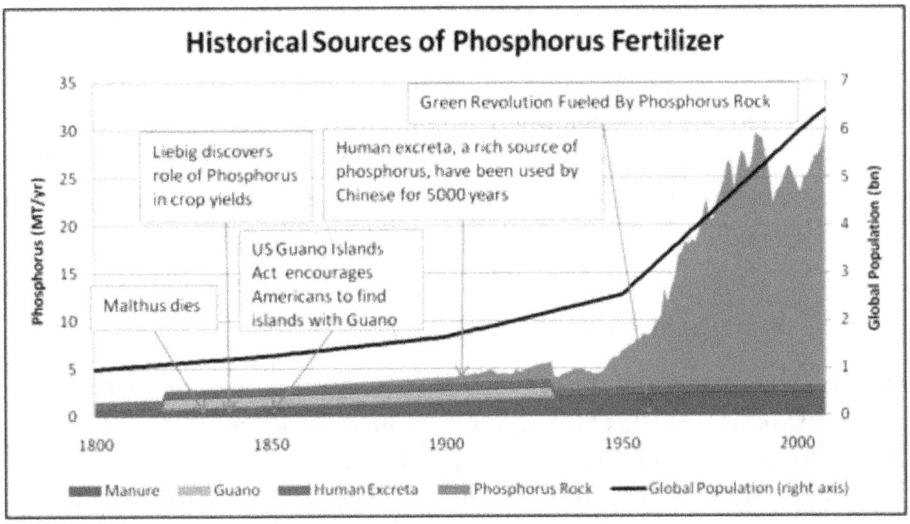

FIGURE 4.3 Historic sources of phosphorus fertilizer and total use over time. Reprinted from *Global Environmental Change*, 19/2, D. Cordell, J.-O. Drangert, S. White, The Story of Phosphorus: Global Food Security and Food for Thought, 292–305, Copyright (2009), with permission from Elsevier.

When considering phosphorus mining holistically, it appears to be more environmentally friendly to recover nutrients from wastewater than to mine new sources. Recovered sources can be generated close to the point of use, reducing the effects of transporting fertilizer products from far reaches of the globe. And, finally, the supply line (e.g., treatment facility influent) is ever present.

The Haber-Bosch process, which converts atmospheric nitrogen to ammonia, ammonium nitrate, and urea for use as fertilizer, if often cited as one of the most important discoveries of the 20th century, as this led to more efficient farming practices (Smil, 2004). However, it is estimated that the process consumes 3 to 5% of the world's natural gas production (about 1 to 2% of the world's energy supply) (International Fertilizer Industry Association, www.fertilizer.org). In addition, the process has affected the natural nitrogen cycle. Runoff from the use of manufactured fertilizers has substantially increased nitrogen in the nation's waterbodies, resulting in anoxic conditions in the Gulf of Mexico, Chesapeake Bay, and Long Island Sound, to name a few. Nitrogen recovery from biosolids in the form of land application, compost, or fertilizer production potentially provides a more sustainable means of providing nutrients for agriculture. Processes are beginning to be developed to make wastewater a cost-effective, sustainable resource for fertilizer production. These products are slow-release products, which are also more sustainable, as nutrients become available when most needed for plant growth.

On the horizon is the capture and use of urine, a relatively sterile stream, as a nutrient source. Urine contributes approximately 80% of total nitrogen, 55% of total phosphorus, and 60% of potassium in domestic wastewater. Keeping this stream as a separate stream eliminates much of the preprocessing and captures the nutrients at the source. Although likely not viable where a vast network of infrastructure exists, new developments could consider this as the next generation of wastewater treatment in developing countries.

1.3 Economics of Resource Recovery

Each treatment facility is different in its size, location, receiving water characteristics, permit requirements, and treatment processes used. Therefore, the economics around treatment and resource recovery are also unique. Resource recovery can be thought of in three overarching buckets: energy recovery (heat and power), biosolids and nutrients (fuel and fertilizer), and reclaimed water. As aging treatment facilities (many of which were constructed at the advent of the Clean Water Act) around the nation look to expand and/or meet more stringent effluent requirements, upgrading the facility provides a prime opportunity to look at the facility holistically and to ask the following questions: "What can and should this facility look like in the future? Can the facility become 'net zero' in terms of energy and water? And will the economy foster this change as resources diminish?"

In addition, questions for utility managers to consider when assessing economics include the following:

- Is an individual or regional solution more cost-effective based on facility size?
- Is there a market locally for the resources produced (i.e., biosolids, water, energy, and fertilizer)?
- How does the cost of power in the region influence the decision?
- How does climate change (e.g., droughts, floods, and sea level rise) influence the decision?
- What can the utility do to reduce power consumption to save costs and reduce greenhouse gas emissions?
- What is the value of potable water in the region?
- How will the cost of fertilizer affect decisions?
- Are there revenue streams (septage, sludge, waste organics, food, and beverage processing wastes) that the utility should investigate? and
- Will regulations on the horizon influence economics?

This chapter will expand on the water–energy–nutrient nexus, discuss how these factors are interrelated and how the economics factor in, and acknowledge that each facility is different and, therefore, will have a different solution.

2.0 BENEFICIAL USE OF BIOSOLIDS

2.1 Introduction and History of Biosolids Management

As presented in Figure 4.3, manure, guano, and human excreta have been used as a soil amendment and fertilizer for centuries, even millennia. Originally, land application of wastewater sludge was used to sustainably improve and maintain productive soils and to stimulate plant growth. As the population increased, so did the quantities of wastewater sludge to be managed, and sludge became a waste product to be disposed of as cost-effectively as possible. As a result, in addition to land application, ocean dumping (until its ban in 1988), landfilling, and incineration (to reduce volume) became popular means to dispose of these solids.

In 1993 in the United States, the 40 CFR Part 503 Rule was promulgated to regulate the land application of biosolids as a soil amendment. Part 503 established standards on pollutant limits, vector attraction reduction requirements, management practices, and operational standards for land application of biosolids. This changed how many utilities managed their biosolids.

Digestion of wastewater solids (both aerobic and anaerobic) was used originally to reduce the volume of biosolids requiring disposal and for vector attraction and pathogen reduction (the methane gas produced in anaerobic digesters was often simply flared). Today, as the industry recognizes that this methane gas can be harnessed to produce energy, flaring is becoming obsolete.

The Milwaukee Metropolitan Sewerage District was the first utility to produce and market a fertilizer product from wastewater sludge; in fact, it has been doing so since 1925. Other utilities, such as the Massachusetts Water Resources Authority, rely on heat drying of sludge to form a pelletized product to be used as a fertilizer.

Figure 4.4 presents current U.S. biosolids management practices (NACWA, 2010). As shown, more than half of the biosolids generated are beneficially reused; however, as the discussion of sustainability and reclamation becomes more prevalent, this percentage will increase.

2.2 The Future of Biosolids Management

As presented in Figure 4.4, even today, more than one-quarter of the nation's biosolids are landfilled. This represents the greatest opportunity

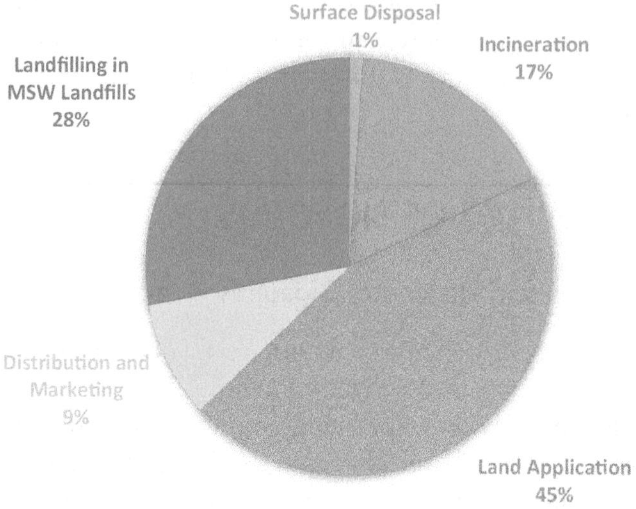

FIGURE 4.4 U.S. biosolids management practices (NACWA, 2010).

for redirecting this waste stream to generate energy. Unprocessed biosolids typically contain about 18 608 kJ/kg (8000 Btu/lb) on a dry-weight basis, which is similar to the energy content of low-grade coal (NACWA, 2010). Diverting this stream from landfills to energy generation will help the nation reduce its dependence on fossil fuels.

As a nation, and as utility managers, the mindset of the industry must shift to view biosolids as a commodity and a resource rather than a waste product. The resource can take the form of a soil amendment, a fertilizer product, and/or a renewable energy source. While many technologies currently exist and are proven in the industry, the industry should continue to push the envelope to improve technologies that exist and participate in the development of new technologies to meet the resource needs of the planet.

As more stringent effluent permit limits are established, especially with respect to nitrogen and phosphorus, the quantity of biosolids will increase (compounded by the increase in solids produced by the expanding population). In addition to the increase in quantity, the characteristics of the sludge can also change. Chemical sludges produced when chemical addition is required for phosphorus removal can be more difficult to thicken and dewater. In addition, as mentioned, there is no measurable energy value in the inert chemical in the sludge. As phosphorus limits go down, chemical addition goes up. This could direct those in the industry to solutions that minimize or eliminate chemical addition as an alternative for phosphorus removal.

2.3 Considerations for Resource Recovery with Biosolids

2.3.1 Facility Size

Facility size can drive decisions regarding resource recovery because often there is an economy of scale associated with processes that recover energy and/or nutrients. For small facilities, less than 19 ML/d (5 mgd), it is often more cost-effective to transport thickened sludge to a larger regional facility for further processing rather than processing on-site. Small utilities should look to partner with neighboring communities or counties to pool resources to increase the economy of scale, or look to a public–private partnership with a regional residuals management facility that is in the business to make the economics favorable. The vision is that as more regional facilities are constructed and prove successful, sludge will move toward becoming a commodity to provide a consistent feedstock to these energy-production or fertilizer-producing facilities. Costs for disposal will become more competitive as these "for-profit" facilities look to attract new sources and generate more product (revenue).

2.3.2 Facility Location

Opportunities for resource recovery are also driven by the location of the facility. In rural areas, opportunities for land application of Class A or Class B biosolids are much more prevalent and cost-effective than in urban areas where transportation costs drive the economics. Smaller facilities may have an outlet for land application that is cost-effective and does not require the sophistication of a digester facility. In rural areas, where land is plentiful, composting of biosolids could be the most cost-effective solution. Facility location, in terms of climate, may also influence the decision. In more temperate regions, the opportunity for land application of biosolids and fertilizers can be a year-round proposition, whereas, in colder regions of the country, the facility may need to incorporate a significant volume of storage to the plan.

2.3.3 Existing Infrastructure

The existing infrastructure in place at the WRRF can dramatically influence the economics of the upgrade. Certainly, if a facility has digesters in place, the cost to upgrade and expand the facility to harness the biogas produced (methane and carbon dioxide [CO_2]) for process heating coupled with the addition of a combined heat and power (CHP) system could have a favorable return on investment. The power generated would reduce the cost of electricity and fossil fuels used at the facility, increase the power reliability at the facility, and provide a renewable fuel that could result in rebates or credits from the local power company.

Facilities that currently incinerate sludge and meet air quality standards should look to capture and use the heat off the system, but keep an eye toward the future. What is the expected life of the incinerators and what is the future of sludge management? When considering the economics of the system, it is important to understand the fossil fuels required to operate the incinerators and, if moving toward chemical phosphorus removal, how this will affect fuel consumption and incinerator performance. It is also important to include in the analysis the cost of air emissions control and to understand how new air regulations may affect the capital and operational costs of maintaining the incinerators.

2.3.4 Regional Economics

The cost of power across the continental United States ranges from about \$0.0713 to \$0.1735/kWh (see http://www.eia.gov/electricity/monthly/epm_table_grapher). Certainly, the economics of power generation, or the return on investment, at a treatment facility will be influenced by power cost in the region. However, the analysis should also include the availability of potential revenue streams from accepting outside waste (e.g., organics, FOG, and residuals from other treatment facilities). In addition to the revenue from accepting outside waste, in the case of FOG and organics, accepting this waste stream can also increase the energy production of the digesters (Parry and Evans, 2012).

The economics of the disposal of biosolids in the agricultural market can vary based on supply and demand in the region. If or when supply exceeds demand, utilities must have alternative disposal options.

2.3.5 Regulations

Existing and potential state and federal regulations should always be monitored because these regulations could affect the decision-making process. This includes water quality regulations as the digester sidestream can affect the ability of the treatment process to meet stringent nutrient limits, air regulations that may make incineration cost-prohibitive, and solid waste regulations that could result in a revenue stream that does not currently exist. For example, a new regulation in Massachusetts bans landfilling of organics for businesses and institutions that generate more than 0.9 metric ton (1 ton) per week of food waste. Generators are now looking for a cost-effective means to dispose of this waste. Digesters at WRRFs can be an attractive outlet.

2.4 Options for Resource Recovery of Biosolids

2.4.1 Land Application

As presented in Figure 4.4, the majority of biosolids in the United States are land applied because these biosolids, rich in nutrients, improve the quality of soil and stimulate plant growth. Biosolids are used in both residential and commercial landscaping, in agriculture, and to reclaim mining sites. Land application of biosolids takes place in all 50 states. As discussed, the Part 503 regulations establish the guidelines for land application of biosolids. Treatment processes used before land application include digestion, composting, and lime stabilization.

Nutrients found in biosolids, such as nitrogen, phosphorus, and potassium and trace elements such as calcium, copper, iron, magnesium, manganese, sulfur, and zinc are necessary for crop production and growth. The use of biosolids in agriculture reduces the farmer's production costs and replenishes the organic matter that has been depleted over time. The organic matter also improves soil structure by increasing the soil's ability to absorb and store moisture. In addition, where water availability is an issue, biosolids can add some level of moisture to the soil.

The organic nitrogen and phosphorus found in biosolids are used efficiently by crops because these plant nutrients are released slowly throughout the growing season. This enables the crop to absorb these nutrients as it grows. This efficiency lessens the likelihood of groundwater pollution of nitrogen and phosphorus, which is a synergy that provides a win for both land and water.

If moving in this direction, it is important for a utility owner to understand and consider agronomic rates of application of biosolids and the need to incorporate a facility to store biosolids on-site. It is also important to acknowledge the restrictions and risks associated with the presence of heavy metals and other regulated pollutants in the biosolids.

2.4.2 Digestion

Anaerobic digestion converts readily biodegradable organics in the biosolids to biogas (methane and carbon dioxide) in the absence of oxygen. The biogas can be used in the process to heat the digesters and also, in conjunction with a gas processing system, can produce power for the facility through a CHP facility. The heat generated from the power generation system can also be recovered and used to heat or cool the process buildings. The quantity of biosolids requiring disposal is reduced by about 40% in the digestion process, and these solids can be land applied or used in land reclamation. Anaerobic digestion is a proven technology, and the systems being used to recover the biogas have advanced over the years. Anaerobic digestion is typically best suited for facilities that use primary clarification and facilities with an average daily flow greater than 19 to 37 ML/d (5 to 10 mgd), although digestion has been used successfully at smaller facilities.

Introduction of organics and FOG to the digestion process can provide a revenue stream for the utility while increasing gas production. This is another example of how the utility can turn waste products into commodities. The FOG in the collection system causes sewer backups and, once at the treatment facility, can generate odors and foul instruments and equipment. Collecting FOG at its source and introducing it to the digester directly can reduce operation and maintenance issues associated with FOG while reportedly improving the digestion of the wastewater solids (another win-win proposition).

Pretreatment of solids before digestion can improve the digestibility of the solids. Thickening of the solids before digestion can significantly increase the digester capacity and/or improve solids destruction because of the increased solids retention time. A number of emerging technologies have looked into cell lysis (i.e., breaking the cell walls) before digestion, including thermal hydrolysis (Cambi, Biothelys, and Exelys), mechanical disintegration (Crown Biogest and Microsludge), and electrical pulse treatment (OpenCEL). It is estimated that biogas production can be increased by 30 to 60% with pretreatment. In October 2014, DC Water became the first facility in the United States to incorporate thermal hydrolysis (using the Cambi Process) ahead of anaerobic digestion at the 1400-ML/d (370-mgd) Blue Plains Advanced Wastewater Treatment Plant. As part of a comprehensive biosolids program, DC Water also constructed a CHP facility, and commissioning is scheduled for late 2015. These projects will result in a \$20 million savings in operation and maintenance costs (approximately \$10 million savings from the reduced quantity of stabilized biosolids to dispose of and \$10 million in energy savings from power production).

2.4.3 Incineration

With an estimated 17% of biosolids generated in the United States incinerated, it is worth noting the potential for resource recovery that exists. In this process, biosolids are burned in a combustion chamber with oxygen, forming carbon dioxide and water and producing ash. Air emissions from the incinerators must be treated to meet state and federal guidelines. Either multiple hearth furnaces or fluidized bed incinerators are used at WRRFs. The benefits of incineration include a significant reduction in the byproduct that must be disposed of (i.e., ash). In a well-run incineration facility, burning can become autogenous (e.g., no supplemental fuel is required to maintain the burn). As with digestion, adding FOG directly to the incineration process can be a better location to introduce FOG to the treatment system as a means to reduce the supplemental fuel required to maintain the burn and to reduce other operation and maintenance issues in the collection and treatment system, as long as regulations allow for this. Heat can be recovered from the incineration process and used to heat process buildings or to generate power. Although there is not yet an efficient process to remove nutrients from the incinerator ash, studies are ongoing, and some facilities are intentionally landfilling ash in a mono-fill for future mining if and when nutrient recovery from ash becomes a profitable venture.

2.4.4 Other Thermal Processes

Gasification and pyrolysis are two other thermal processes that are relatively new in the wastewater industry that will need to be proven over time. With gasification, biosolids are first dried to 50 to 90% and dried solids are fed to a gasifier that produces a syngas, which can be used for heat and power. Additional information about these processes can be found in *Enabling the Future: Advancing Resource Recovery from Biosolids* (WEF, 2013).

3.0 NUTRIENT RECOVERY FROM THE TREATMENT PROCESS

3.1 Introduction

Typically, the influent to a municipal WRRF contains 10 to 50 mg/L of nitrogen and 1 to 10 mg/L of phosphorus. Through the treatment process, nutrients are removed in the biosolids, in the gaseous state (in the case of nitrogen), or discharged in the effluent. Nutrient recovery from the treatment process can be in the form of the liquid stream (nutrients in the facility effluent used for irrigation), in the biosolids (in land application), or as a chemical nutrient product. This section will focus on the chemical nutrient products.

3.2 Sidestream Treatment

Nutrient recovery is most effective when nutrient concentrations are as high as possible, typically above 1000 mg/L total nitrogen and 100 mg/L total phosphorus and when flows are relatively low (WEF, 2013). As such, the most attractive location for nutrient recovery is in sidestream treatment, especially when anaerobic digestion and biological phosphorus removal are incorporated to the treatment train. Sidestreams produced from anaerobic digestion typically contain 15 to 30% of the total nutrient load in the facility and, therefore, constitute a significant source of nutrient recovery. This aligns with the overall goal of resource recovery as anaerobic digestion has the ability to produce energy through the capture of methane gas, as discussed previously. The added benefit resulting from recovery of nutrients from the sidestream is the reduced nutrient load returned to the head of the facility and associated reduction in operational costs (in terms of power and chemicals) required to treat and remove this load. In addition, the operation and maintenance issues associated with struvite formation in internal piping can be minimized. As nutrient limits in the effluent become more stringent, the opportunity for nutrient recovery increases.

When assessing nutrient recovery, the economic evaluation must include the revenue from the sale of the product generated and how its value will likely increase over time as the cost of conventional fertilizers increases. Fertilizer prices have been steadily increasing as power costs increase (required for the conversion of nitrogen gas to nitrogen-based fertilizer products) and supplies of phosphorus rock decrease. As the cost of conventional fertilizers continues to increase and the cost to extract nutrients from the waste stream decreases, the balance will tip in favor of nutrient recovery at the WRRF.

3.2.1 Struvite Precipitation

Struvite (magnesium ammonium phosphate) has long been the bane of many treatment facility operators who used anaerobic digestion in their treatment train. The digestion process, in combination with the right concentrations of magnesium, ammonia, and phosphates under a high-pH environment, results in the precipitation of struvite in pumps, piping, and equipment, compromising the efficiency of the system and often requiring the complete replacement of systems. Struvite recovery processes such as Ostara's Pearl® process has capitalized on this challenge and turned it into an opportunity. These processes capture phosphorus and nitrogen from municipal and industrial waste streams and transform them, in a controlled environment, into a potentially marketable slow-release fertilizer. As discussed, these processes are most cost-effective with the high sidestream concentrations found with anaerobic digestion and biological phosphorus removal.

The Metropolitan Water Reclamation District of Chicago (Illinois) will begin recovering phosphorus and nitrogen from its Stickney Water Reclamation Plant in the fall of 2015 using the Ostara technology. Once fully operational, the facility will be the largest of its kind in the world, with the potential capacity to produce between 10 000 to 15 000 tons of struvite crystal fertilizer annually. The resulting fertilizer, marketed as Crystal Green®, further reduces nutrient loss because of leaching and runoff because it is highly water insoluble, and instead releases nutrients in response to plant demand. Other facilities using the Ostara process include the Rock Creek Advanced Wastewater Treatment facility in Hillsboro, Oregon, and the Edmonton Gold Bar Wastewater Treatment Facility is Alberta, British Columbia, Canada.

Other technology providers based on struvite recovery include Multiform Harvest, NuReSys®, Phospaq™, CNP Airprex®, and Crystalactor®.

3.2.2 Waste Activated Sludge Treatment

For facilities that use biological phosphorus removal, the waste activated sludge (WAS) contains higher concentrations of phosphorus than conventional WAS that can be released by creating anoxic or anaerobic conditions. The WASStrip™ process developed by Clean Water Services creates these conditions in a controlled environment and has been used upstream of WAS thickening and digestion to concentrate phosphorus and magnesium in the liquid phase to reduce the production of struvite in the system. Paired with the Ostara recovery system, the phosphorus available for recovery is doubled and the amount of supplemental magnesium required in the process is reduced.

3.2.3 Ammonia Removal

Although not a nitrogen recovery process, sustainable means to reduce the ammonia load from a high-strength sidestream exist. One such process, the DEMON® process, is a nitrification/deammonification process in which ammonia and nitrite are simultaneously converted to nitrogen gas, without the use of organic carbon. The anaerobic ammonium oxidation (anammox) process is being investigated as a sustainable ammonium removal technology. Because the anammox bacteria convert ammonium and nitrite directly to nitrogen gas anaerobically, the oxygen demand is significantly reduced.

3.3 Fertilizer Manufacturing

As discussed previously, production of fertilizer from biosolids has been practiced since the 1920s. However, many such technologies have not had success because of product quality, odors, and transportation costs. Newer technologies have merged biosolids into the fertilizer manufacturing process to generate a product that looks and feels more like a fertilizer than biosolids One such product is VitAG fertilizer. The VitAG fertilizer mixes dewatered biosolids (18 to 30% solids) with ammonia and concentrated phosphoric and sulfuric acids in a hydrolysis vessel, conditions the material with additional nutrients and binding agents, and then processes the material through a granulator and dryer system, followed by further product screening and coating. The benefit of this process is that it harnesses the nitrogen in the biosolids in the product. Other companies that have plans to, or currently produce, organically enhanced inorganic fertilizers are Green Technologies, Inc., Harmony Products, Inc., BER Systems, Inc., and Unity Envirotech, LLC.

3.4 Algae Treatment

The need for nutrient removal from a WRRF stems from cultural eutrophication in the receiving waters. In recent years, many have studied the potential for harnessing the algae in a controlled environment and turning the problem into the solution. The benefits provided by algae are that it removes nutrients from the liquid stream, without chemicals, to extremely low levels; sequesters CO_2; and produces O_2. And harvested algae will likely have many applications as a beneficial product.

The challenges of algae treatment are twofold: the large area required for a pond system (because the penetration of sunlight is critical to algae growth) and the separation of algae from the clean water. The Clearas Water Recovery technology has worked to manage these two challenges. Similar to an activated sludge system, the Clearas system cultivates algae (versus bacteria) in a controlled and continuous-flow environment in the photobioreactor and recycles algae to maintain a viable population. Process control is managed based on retention time, nutrient load, and biomass in the system. The photobioreactor is comprised of clear tubing that optimizes biological activity and photosynthesis. Nutrients and CO_2 are consumed and oxygen and new algae cells are produced. In a separation stage, membranes are used to separate the algae from the clean water. Surplus algae are removed to keep the ratio of biomass and food balanced. The waste algae, high in lipid content and nutrients, is a beneficial product that can be used for a variety of purposes, including fertilizers and feedstock. The system has been successfully piloted at an industrial application and a small (primarily residential) treatment facility in the Pacific Northwest. In addition, piloting is currently ongoing at the Upper Blackstone Water Pollution Abatement District serving Worcester, Massachusetts. Other researchers in Europe are experimenting with high rate algae ponds as a means to treat anaerobically digested urban wastewater.

4.0 WATER REUSE

4.1 Introduction

In terms of water reclamation, treatment facilities located in and around drought-stricken areas are moving quickly to provide the treatment necessary to produce reclaimed water to supplement water supply, sometimes regardless of cost, but out of necessity. Similarly, in areas where sea level rise could threaten water supplies, reclaimed water is being injected to aquifers to prevent saltwater intrusion, again driven more by necessity and the cost of developing an alternative water supply.

In areas where stringent nutrient standards may drive treatment process selection, reclamation should be considered, even in areas that are considered "water rich". Because many of the processes used to achieve the strict phosphorus standards meet California Title 22 standards for reclamation (e.g., membranes and filtration), the evaluation of alternatives should address the potential cost–benefits of incorporating reuse to the facility upgrade.

In all cases, utility owners need to understand their service area, know the large water users in the system, and consider means to provide an alternative water source to these users, which may include agriculture, industry, power plants, and irrigation.

U.S. Environmental Protection Agency's (2012) *Guidelines for Water Reuse* presents reuse water quality requirements, which vary by state and ultimate use, and a number of successful case studies. This resource should be referenced when considering reuse options. It is important for the utility owner to understand that there may be conflicting opinions regarding removing effluent from a receiving water for reuse given the effect to stream flow, especially in drought-prone areas. In addition, water rights issues, specifically in the Southwest, are critical and cannot be overlooked.

4.2 Potable Reuse

Direct and indirect potable reuse is becoming more prevalent, more quickly than expected, as the effects of climate change become a reality. Utilities have been forced to assess alternative sources of water in drought-stricken areas, specifically in California, Texas, and the Southwest, where utilities are also taxed with increases in population at a higher rate than the remainder of the nation. The discharge from the WRRF can quickly become the most viable, and cost-effective source when compared to desalination, significant conveyance projects, and/or treatment of less desirable sources. Again, this is an area where utility owners must look at their system holistically and consider the following questions:

- What is the current source of potable water (and how resilient is it to climate change)?
- What is the quality of this source and what does it cost to treat it?

- Who are the largest users in the system and do they require a potable source?
- What NPDES discharge limits do you currently need to meet and what are they expected to be in the future?
- What more would you have to do to produce reuse-quality water at the WRRF?

By having a holistic understanding of the water cycle in your community and/or your watershed, opportunities come to light that may heretofore have been in the shadows. One utility owner conveyed a story where reuse-quality treatment facility effluent was discharged into a contaminated aquifer only to be pumped out and treated for potable use. By taking the aquifer out of the flow scheme and implementing a direct connection from the WRRF effluent to the water treatment facility influent, the cost of pumping and treatment could be reduced significantly. Certainly, there are public perception hurdles to cross, but, by knowing and accurately explaining the alternatives to the public, more sustainable solutions can arise. One community in the Mississippi basin, after realizing the source of their potable water was the Mississippi River, which contained wastewater effluent from numerous upstream communities, decided they would rather drink their own effluent than effluent from someone else.

4.3 Irrigation

Irrigation of crops, both for human consumption and nonhuman consumption, represents another significant demand on the water resources of the nation. Although advances have been and continue to be made in water conservation in the agricultural industry, continued collaboration between utility owners and the agricultural industry needs to be fostered, not only in terms of water, but in terms of nutrients and energy. By forming new partnerships with farmers, utility owners can better understand the symbiotic relationship between the two entities. Indeed, utility owners often regard the agricultural industry as the source of nutrients contaminating the waterways of the nation, and feel that regulations aimed at reducing nutrients from WRRFs are misdirected. And, certainly, the agricultural industry needs to acknowledge and manage (and perhaps capture and reuse) nutrients in the runoff from farmlands. Regulators working in collaboration with utility owners also need to determine whether it is better to leave the nutrients in the treated effluent if reused for irrigation (rather than expending the energy to remove the nutrients from the liquid stream, only to extract them from the solids stream with more energy). As discussed previously, land application of biosolids plays a role in this discussion.

4.4 Water for Energy Production

As mentioned, the production of electric power represents one of the largest uses of water in the nation. Oftentimes, especially in urban areas, a WRRF is located in close proximity to a power plant. This provides an opportunity for both entities. The improved effluent quality, driven by more stringent nutrient limits, could be of sufficient quality for cooling water at the power plant (although perhaps not of reuse quality). The power plant could benefit by not having to purchase potable water and the WRRF could benefit from reducing the nutrient load into receiving waters. A warm sidestream sent back to the WRRF from the power plant could provide benefits to the treatment facility by increasing the wastewater temperature, which can foster nitrogen removal while addressing any temperature limitations on the effluent. This represents a specific market sector where it could be cost-effective to use reclaimed water and maintain potable water supply for true potable uses.

If the effluent source is not close to the power production facility, then consideration of a "scalping facility" to treat the wastewater to the level needed (i.e., where it is needed at the point of demand) can help a system move toward carbon neutrality.

4.5 Effect of Effluent Reuse on Total Maximum Daily Load

Again, it is important that all stakeholders are involved in the discussion of reuse. Redirecting facility effluent that has been a significant source of flow in a receiving water to reuse may create concerns among river advocates. The receiving waters need to be fully understood in terms of quantity and quality of water being discharged and the effect to the receiving water if this source is removed (both from a quantity and quality perspective).

5.0 OTHER RESOURCE RECOVERY

5.1 Metals

Historically, metals have been removed from a waste stream because of environmental regulations; however, as the price of metals increases, the trend is toward recovery. At this point, this practice could be most economical at facilities that have a higher concentration of metals such as in the mining or metal plating industry. Companies that incinerate municipal solid waste are looking into metals recovery from incinerator ash. Although recovering metals from a municipal WRRF may currently be in its infancy, as technologies advance there could be a market in the public sector.

5.2 Chemical Recovery

Similar to metals, certain chemical removal technologies are best suited for industrial process streams where the flows are lower and the chemicals are more concentrated. Companies such as ThermoEnergy are in the business of providing sustainable solutions for chemical recovery from various manufacturing processes using a vacuum distillation process.

5.3 Heat

Regardless of the region, wastewater typically remains at a relatively constant temperature as it is conveyed to the treatment facility and processed on-site. The latent heat in the wastewater can be used with heat pumps to generate power or provide heating and cooling for buildings. This has been used successfully with several treatment facility effluents in New England. Although the offset is not significant, often it is the number of smaller sustainable projects that add up to the larger sum. The wastewater can also serve as a heat sink, if needed, as long as effluent temperature requirements are met.

5.4 Alternative Carbon Sources (for Nutrient Removal)

When total nitrogen limits are required to be reduced to 5 mg/L or less, typically a supplemental carbon source is required in the treatment scheme to achieve the low levels. The need to add carbon back into the process seems counterintuitive because energy has already been spent removing carbon from the system. Alternatives to methanol are synthetic products (such as MicroC™), which are safer to handle, but can be expensive. Along the lines of resource recovery, the utility owner should look into byproducts in other industries in and around the service area as a source of carbon for nutrient removal. Industries that show potential include breweries, soft drink manufacturers, and food producers. This waste could currently be entering the treatment system through the sewer line or as septage; however, if stored and introduced to the treatment process at the right location and in the right concentrations, it could eliminate the need for an outside carbon source. It is important to understand that this process is not a panacea, and would have to be piloted to assess its viability. It is also important to note that when working with waste from the food and beverage industry, oftentimes the purity of source can vary based on the production line; as such, the consistency of the product can be difficult to manage.

The production of volatile fatty acids through primary sludge fermentation can also assist in the biological nutrient removal process by harnessing and returning the available carbon to the most beneficial point in the system.

6.0 REFERENCES

Cordell, D.; Drangert, J.-O.; White, S. (2009) The Story of Phosphorus: Global Food Security and Food for Thought. *Global Environ.* Change, 19, 292–305.

International Fertilizer Industry Association. Fertilizer Statistics. www.fertilizer.org (accessed Feb 2015).

Jasinski, S. M. (2012) *Mineral Commodities Summaries: Phosphate Rock*; U.S. Geological Survey: Reston, Virginia.

Massachusetts Institute of Technology. Mission 2016: The Future of Strategic National Resources, Phosphorus: Supply and Demand. http://web.mit.edu/12.000/www/m2016/finalwebsite/problems/phosphorus.html (accessed Feb 2015).

National Association of Clean Water Agencies (2010) *Renewable Energy Resources: Banking on Biosolids;* National Association of Clean Water Agencies: Washington, D.C.

Parry, D. L.; Evans, P. (2012). Wastewater and Organic Waste to Bioenergy. *Water Practice Technol.,* 7 (4).

Smil, V. (2004) *Enriching the Earth: Fritz Haber, Carl Bosch, and the Transformation of World Food Production*; MIT Press: Cambridge, Massachusetts.

U.S. Environmental Protection Agency's (2012) *Guidelines for Water Reuse;* EPA-600/R-12-618; U.S. Environmental Protection Agency: Washington, D.C.

Water Environment Federation (2013) *Enabling the Future: Advancing Resource Recovery from Biosolids;* Water Environment Federation: Alexandria, Virginia.

5

Alternatives Evaluation

1.0 INTRODUCTION

This chapter provides a comprehensive outline, with additional details, on how to execute an upgrade to a wastewater treatment system to accomplish nitrogen and phosphorus removal. The purpose of this chapter is to provide guidance to managers and engineers on process options and configurations to achieve nutrient removal from wastewater in compliance with increasingly stringent National Pollutant Discharge Elimination System (NPDES) permits. Guidance is also provided to assist in establishing objectives to plan, design, and operate municipal wastewater treatment facilities to remove nitrogen and phosphorus from reclaimed water.

2.0 OVERVIEW OF APPROACHES AND TECHNOLOGIES FOR CONTROLLING NUTRIENTS FROM NONPOINT (DIFFUSE) SOURCES

2.1 Nonpoint Sources of Nutrients

Nonpoint sources (NPSs) of nutrients originate from many diffuse sources. As water originating as rainfall or snowmelt travels over ground or through shallow groundwater, there is an accumulation of nutrients from natural and anthropogenic areas within watersheds. Nutrients are eventually transported to receiving inland or coastal waters. Manmade features such as agricultural drainage ways and municipal stormwater systems provide collection and discharge of irrigation or natural runoff and transport water (and nutrients) to receiving waters. Table 5.1 presents various land uses/activities and the associated sources of nutrients.

The magnitude of the various NPSs of nutrients and the effect of these sources on receiving waterbodies depend on watershed characteristics, transport pathways, attenuation mechanisms, and receiving water characteristics. Watershed-specific studies and assessments need to be conducted to assess the magnitude and importance of various NPSs of nutrients. Munn et al. (2009)

TABLE 5.1 Sources of nutrients in watersheds.

Land use	Nutrient source
Agriculture	Fertilizers, herbicides, insecticides, sediment, irrigation water
Urban/suburban	Fertilizers used on residential lawns, parks, gardens, and golf courses; pet waste, sediment/soils from construction sites, irrigation water, hydromodification
Forested	Naturally occurring fertilizers in timber production areas
Livestock	Animal manure
Septic systems[a]	Septic system leachate, failing septic systems
Atmospheric deposition[b]	Wet, dry, or fog deposition

[a]Septic systems are not a land use, but are often used in rural areas.
[b]Atmospheric deposition is not a land use, but provides a source of nutrients.

reported on the percent of phosphorus and nitrogen loads from farm fertilizer, manure, nonfarm fertilizer, and atmospheric deposition associated with three types of land use. Figure 5.1 presents the results of this analysis and illustrates that highly urbanized watersheds and highly agricultural watersheds, for example, would be dominated by different sources of nutrients.

It is critical to understand nutrient sources within a watershed to implement NPS control measures effectively. There are numerous methodologies for quantifying NPS nutrient loads, ranging from relatively simple empirical/statistical models to more complex mechanistic models. Nutrient loads can also be estimated from monitoring data. There are many documents that outline methods and approaches for estimating nutrient loads in watersheds. U.S. Environmental Protection Agency (U.S. EPA, 2003) outlines various approaches for estimating loads. Borah and Bera (2013) describe various types of models and the mathematical basis. In addition, U.S. EPA websites provide information on public access models (U.S. EPA, 2013b, 2015).

2.2 Approaches and Technologies

Non point-source pollution encompasses a wide range of activities across the landscape that, although individually small, are numerous and difficult to identify and control (NCDER, 2014).

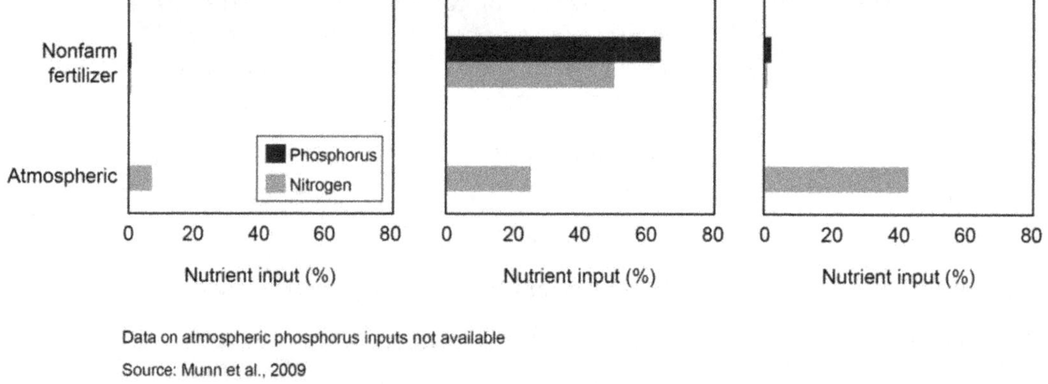

FIGURE 5.1 Phosphorus and nitrogen loads from farm fertilizer, manure, nonfarm fertilizer, and atmospheric deposition associated with three types of land use.

Management practices control the delivery of NPS pollutants to receiving water resources by

- Minimizing pollutants available (source reduction);
- Retarding the transport or delivery of pollutants, either by reducing water transported, and thus the amount of the pollutant transported, or through deposition of the pollutant; and
- Remediating or intercepting the pollutant before or after it is delivered to the water resource through chemical or biological transformation (U.S. EPA, 2003).

Many researchers and agencies have published documents on best management practices (BMPs) for controlling NPS pollution. Some of the most useful documents are those developed by U.S. EPA for the Chesapeake Bay Program. This program focuses on nutrient pollution and addresses a range of urban and agricultural sources for nutrients. While specific to the Chesapeake Bay Watershed, information on control measures for a specific NPS in one watershed (including effectiveness and cost) is useful for other watersheds. Useful U.S. EPA references include the following:

- *National Management Measures to Control Nonpoint Source Pollution from Agriculture* (U.S. EPA, 2003);
- *Handbook for Managing Onsite and Clustered (Decentralized) Wastewater Treatment Systems* (U.S. EPA, 2005a);
- *National Management Measures to Control Nonpoint Source Pollution from Forestry* (U.S. EPA, 2005b);
- *National Management Measures to Control Nonpoint Source Pollution from Urban Areas* (U.S. EPA, 2005c);
- *National Management Measures to Protect and Restore Wetlands and Riparian Areas for the Abatement of Nonpoint Source Pollution* (U.S. EPA, 2005d);
- *Biological Nutrient Removal Processes and Costs* (U.S. EPA, 2007a);
- *National Management Measures to Control Nonpoint Source Pollution from Hydromodification* (U.S. EPA, 2007b); and
- *Guidance for Federal Land Management in the Chesapeake Bay Watershed* (U.S. EPA, 2010a).

Note that each chapter in the last book is organized by types of practices to control nutrients and includes references and links to additional information.

In addition, U.S. EPA's Web page, "Polluted Runoff: Nonpoint Source Pollution" (U.S. EPA, 2014), includes information and links related to various NPSs of pollution and control measures. There is also extensive technical information on BMP performance for a range of pollutants provided in the *International Stormwater BMPs Database* (ASCE et al., 2015).

Implementation of load reduction measures requires site-specific evaluation to determine the type(s) of measures to implement, expected effectiveness, anticipated water quality benefits, and costs. In general, practices can be categorized into source-control measures or treatment measures, and fall under the general terminology of BMPs.

Table 5.2 includes a summary of some BMPs that are applied to reduce nutrient loads originating from various sources. Table 5.2 is not an exhaustive list, and the reader should refer to U.S. EPA's *Guidance for Federal Land Management in the Chesapeake Bay Watershed* (2010a) and the references cited in that document as well as the other publications previously cited.

3.0 OVERVIEW OF POINT-SOURCE NUTRIENT REMOVAL TECHNOLOGIES

Nutrient point sources are typically domestic or industrial/commercial in nature. While the focus of this discussion is on domestic point sources, the technologies mentioned herein or their variations may also be used for industrial applications. A number of technologies are available to remove nutrients from point sources. While most are well-established technologies with many years of successful use, a number of emerging technologies that show promise to improve on current practices are at different stages of development. A recent publication by U.S. EPA (2013a) presents a comprehensive review of emerging technologies, while providing a list of established ones. This section provides a general overview of established point-source nutrient removal technologies. Detailed descriptions of various biological nutrient removal (BNR) technologies can be found in various textbooks and publications from the Water Environment Federation (WEF) and U.S. EPA (e.g., Tchobanoglous et al. [2014], U.S. EPA [2010b], and WEF et al. [2005]).

Biological treatment technologies for nutrient removal (i.e., BNR processes) have been developed over the years to meet specific treatment goals. For example, there are BNR processes specifically designed to remove nitrogen, phosphorus, or both nutrients. Ultimately, BNR systems remove targeted nutrients either through biochemical transformations or by incorporation to new biomass.

Depending on nutrient removal goals, bioreactors are designed to provide appropriate environmental conditions for growth of specific bacteria

TABLE 5.2 Summary of best management practices (BMPs) applied to reduce nutrient loads originating from various sources.

BMP category	BMP type	Description
Agricultural BMPs		
Erosion control	Conservation tillage	Tillage approaches that leave 30% of the field covered with crops and crop residue
	Cover crops	Crops used to cover the ground to prevent erosion and for the purpose of consuming nutrients remaining in the fields
	Crop rotation	Alternating annual and perennial crops
Filtration	Filter Strips	Vegetated areas located near high-pollutant load areas
Livestock/manure management	Animal waste systems	Collection, transportation, and storage of manure
Rotational grazing		Use of grazing animals for vegetation harvest
Nutrient management	Nutrient management planning	Management of nutrient application (fertilizers or manure) to minimize nutrient runoff and leaching
	Enhanced nutrient management planning	Maximizing the efficiency of nutrient use
Nutrient treatment	Riparian forest buffers	Areas of trees, shrubs, and grasses located adjacent to ponds, lakes, and streams that filter out pollutants from runoff and provide shade for fish and wildlife
	Wetland	Constructed or natural wetlands used to treat agricultural wastewaters through biological processes

(*continued*)

TABLE 5.2 Summary of BMPs applied to reduce nutrient loads originating from various sources. (*Continued*)

BMP category	BMP type	Description
Urban BMPS		
	Erosion control	Practices to control or reduce sediment transport (and other pollutants)
	Source reduction	Practices to reduce the source of the pollutant (e.g., residential/commercial/ municipal fertilizer application practices)
	Stormwater treatment—detention	Extended detention basins, tetention/detention ponds
	Stormwater treatment—separators	Hydrodynamic separators
	Stormwater treatment—filtration	Media/sand/compost filters, biofilters
	Stormwater treatment—wetlands	Constructed surface and subsurface wetlands
	Low-impact development/green infrastructure	Maintaining predevelopment hydrology though infiltration, bioretention, biofiltration rainwater harvesting/reuse

Sources: Merriman et al. (2009), U.S. EPA (2010a), and CBC (2004)

populations able to use target organic and nutrient compounds in their metabolism. One of the most important environmental conditions is the presence, or absence, of oxygen or oxidized nitrogen in the bioreactor. When oxygen is present in the form of free dissolved oxygen, the bioreactor is said to be operating under aerobic conditions. When oxygen is present in a combined form, such as nitrate or nitrite, the bioreactor is said to be operating under anoxic conditions. When the presence of oxygen or nitrate/nitrite is negligible, the bioreactor is considered to be anaerobic.

3.1 The Baseline—Conventional Secondary Treatment

Biological treatment technologies are the most cost-effective way to treat domestic wastewater. Simply stated, biological technologies rely on the metabolism of microorganisms to remove organic material and nutrients from wastewater.

In general, biological processes can be divided into three groups, depending on the method for growing the microorganisms used to accomplish treatment. These categories are as follows:

- Suspended growth processes (e.g., activated sludge)—microorganisms grow suspended in the water in the bioreactors. In suspended growth processes, the bacteria tend to naturally clump together into flocs ranging in size from about 10 to 100 μ;
- Attached-growth processes (e.g., trickling or biological filters)—microorganisms grow attached to a surface, forming a film anywhere from 50 to 200 μ thick; and
- Hybrid processes (e.g., IFAS)—processes that use a combination of suspended growth and attached-growth biomasses.

Baseline conventional technologies use biological treatment to remove biochemical oxygen demand (BOD) and total suspended solids from wastewater. This is commonly referred to as *secondary treatment*. Nitrogen and phosphorus removal technologies are generally built on these baseline processes. Simple process flow diagrams for these baseline processes for secondary treatment can be found in Figures 5.2 and 5.3.

3.2 Nitrogen Removal

Nitrogen in domestic wastewater is predominantly in the form of ammonium ion (NH_4^+) and organic nitrogen. The removal of nitrogen from wastewater follows a series of biochemical transformations that naturally occur in the environment as part of the nitrogen cycle. These transformations convert nitrogenous compounds through a series of steps to nitrogen gas, which is returned to the atmosphere.

3.2.1 The Nitrogen Cycle and Biological Nitrogen Removal

The complete nitrogen cycle involves a number of physical and biochemical transformations that convert atmospheric nitrogen to nitrogenous compounds then back to atmospheric nitrogen. The nitrogen cycle pathways, which are critical for nitrogen removal from wastewater, are (1) nitrification and (2) denitrification.

3.2.1.1 Nitrification (Including Nitritation)

Nitrification is the process of biologically oxidizing nitrogen in the form of ammonium ion to nitrate (NO_3). The process is accomplished under

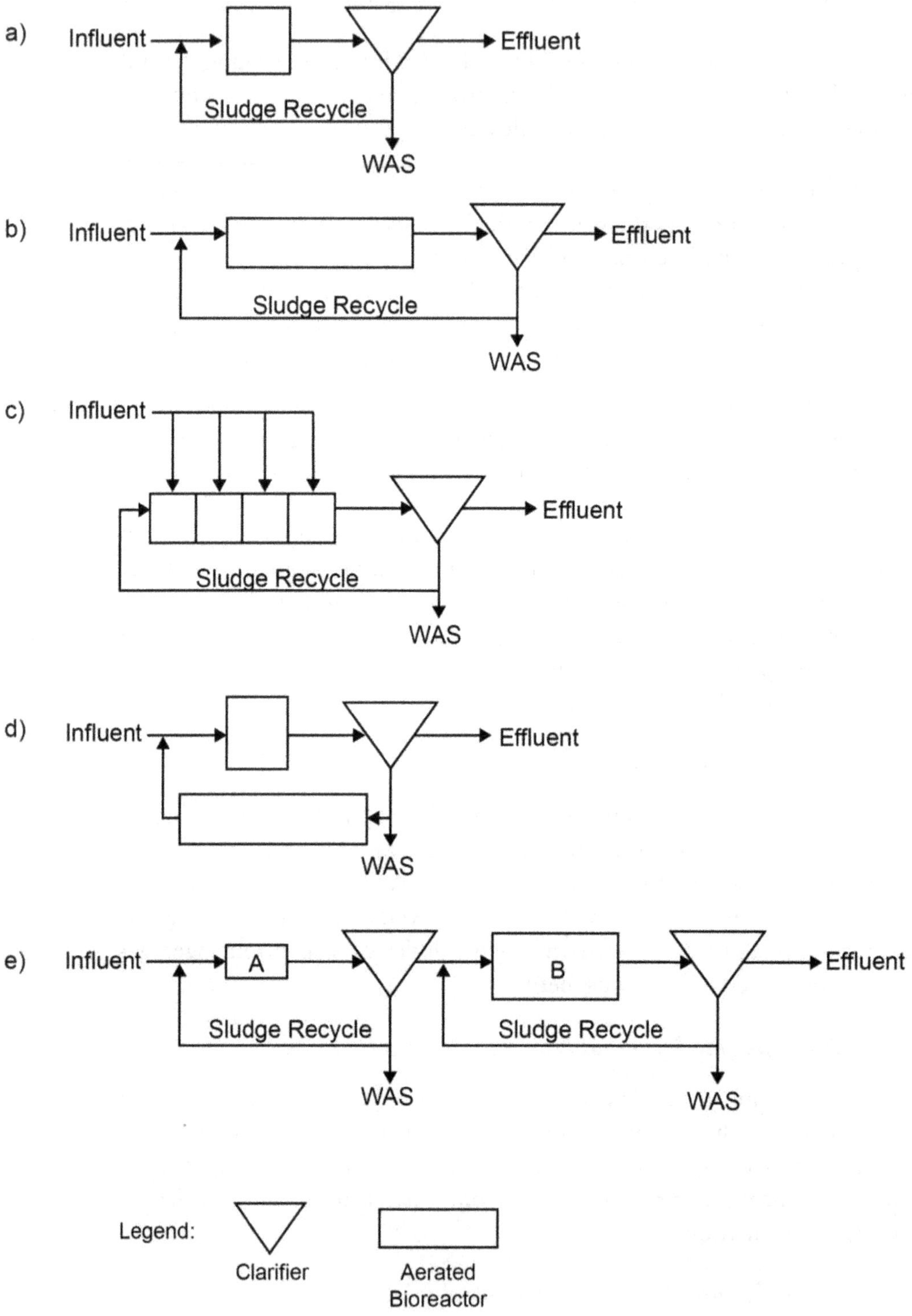

FIGURE 5.2 Suspended growth processes for secondary treatment: (a) complete mix, (b) plug flow, (c) step feed, (d) contact stabilization, and (e) A-B or two sludge.

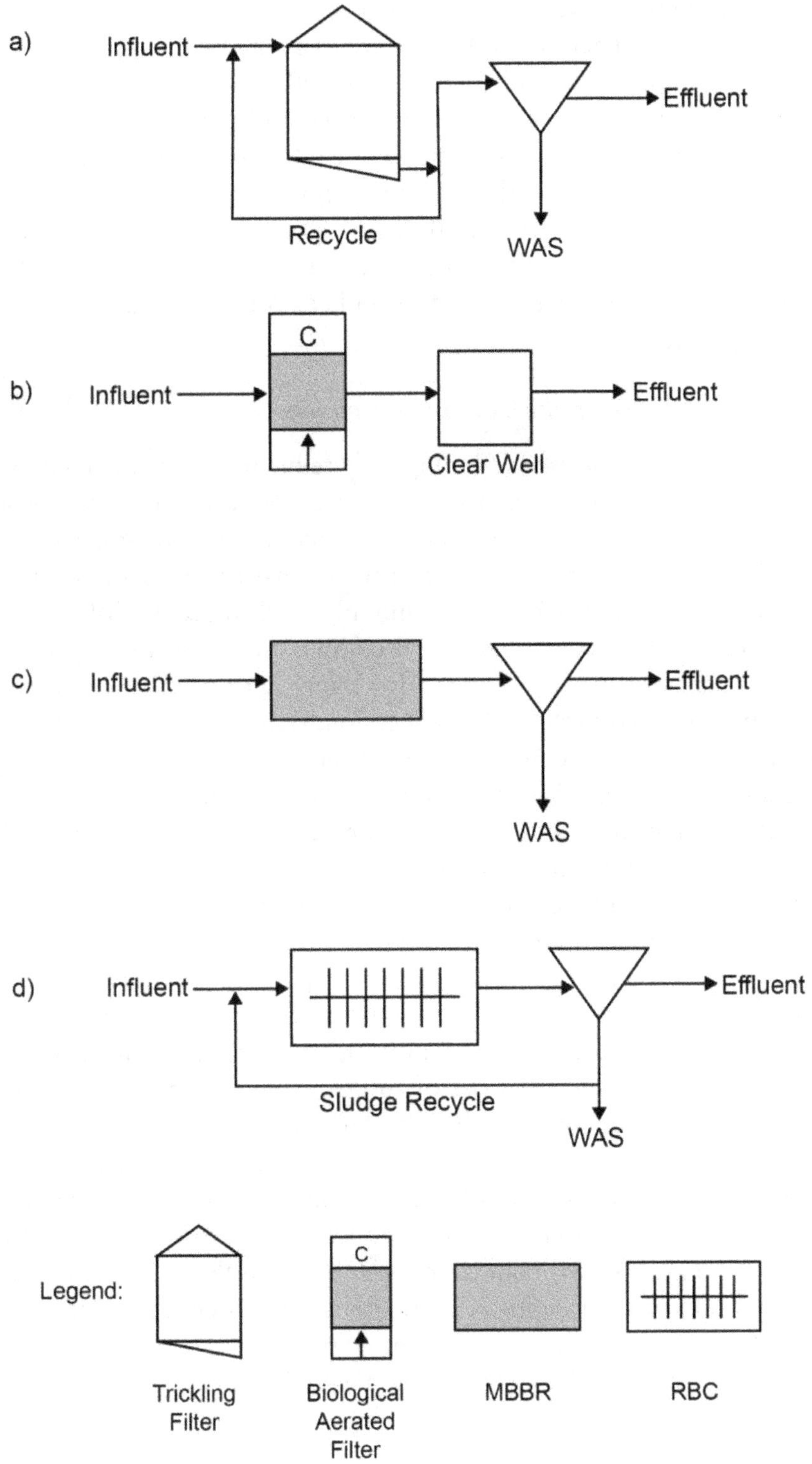

FIGURE 5.3 Attached-growth processes for secondary treatment: (a) trickling filter, (b) biological aerated filter, (c) moving bed biofilm reactors (MBBR), and (d) rotating biological contactor (RBC).

aerobic conditions in two steps by two groups of bacteria. In the first step, ammonia-oxidizing bacteria (AOB) oxidize ammonium ion to nitrite (NO_2); in the second step, nitrite-oxidizing bacteria (NOB) oxidize nitrite to nitrate. The process of nitrification is well established and used to remove ammonia from wastewater. In recent years, interest has arisen in controlling the nitrification process to stop the transformation after the first step, that is, at nitrite (referred to as *nitritation*). Doing so requires process operations to minimize/eliminate the growth of NOB in the system. The benefit is that the amount of oxygen that must be supplied to the biological process is reduced significantly.

3.2.1.2 Denitrification (Including Deammonification)

Denitrification is the process of biologically reducing nitrate through nitrite to nitrous oxides and nitrogen gas. The process is mostly accomplished under anoxic conditions by heterotrophic bacteria that are able to use the oxygen in nitrate or nitrite to oxidize organic matter. The process of denitrification is well established and commonly used to accomplish complete nitrogen removal from wastewater in combination with the nitrification process. A carbon source is required for the growth of the most common types of bacteria that perform denitrification. Depending on effluent limits, the carbon present in raw wastewater can be used to drive denitrification, while, under some conditions, it must be either supplemented or totally provided through the addition of a commercial chemical like an alcohol, glycerin, or sugar. When nitrification is controlled to stop the process at the nitrite level, and then followed by conversion to nitrogen gas, the process is called *denitritation* (i.e., reduction of nitrite to nitrogen gas), and both the oxygen demand and the amount of carbon supplement required are reduced.

Denitrification may also be accomplished by bacteria that do not use carbon (i.e., autotrophic bacteria); however, these bacteria only grow under specific reactor conditions. The use of autotrophic bacteria to remove nitrogen is called *deammonification.* Successful nitrogen removal without carbon requires tight control of reactor conditions to develop an inventory of a specific microbial population. In this case, ammonium ion is oxidized under anaerobic conditions along with the reduction of nitrite to nitrogen gas. These bacteria grow slowly compared to AOB and NOB. To date, application of deammonification in mainstream treatment has been limited to laboratory- and pilot-scale operations, and one or two full-scale research projects.

3.2.1.3 Inhibition

A number of chemicals and environmental conditions can inhibit the metabolism of bacteria involved in the nitrogen removal process. A few examples of compounds that inhibit nitrification include organic solvents, phenolic compounds, cyanide, benzene, un-ionized ammonia and nitrous acid at high concentrations, and metals such as nickel and chromium. In addition, environmental conditions such as pH, temperature, and ammonium and dissolved oxygen concentration significantly affect the growth rate of nitrifiers.

Denitrification is typically inhibited by the presence of oxygen; however, simultaneous nitrification and denitrification (SND) is not uncommon in systems that operate at low dissolved oxygen concentrations. Under some conditions, inhibition of the NOB can result in the accumulation of significant concentrations of nitrite. Residual nitrite is of particular concern to water resource recovery facilities (WRRFs) using chlorine for disinfection. Specifically, 1.0 mg of residual nitrite consumes up to 5.0 mg of chlorine (referred to as *nitrite lock*). High and variable nitrite concentrations will reduce the efficiency of disinfection by chlorine.

3.2.2 Nitrification

Nitrification is done in aerobic bioreactors of varying configurations, as described in the following sections.

3.2.2.1 Suspended Growth Processes

In a single sludge process, wastewater is treated in one aerated tank or bioreactor train where both BOD and ammonia are oxidized. This process configuration is the most common one used for ammonia removal from domestic wastewater without significant industrial contributions.

Two-sludge processes have two activated sludge processes in series. The first is designed strictly for BOD removal, while the second provides nitrification. Historically, two-sludge processes were developed before single-sludge processes based on the belief that each group of bacteria (heterotrophs for BOD removal and nitrifiers for ammonia removal) had to be grown separately. Experience demonstrated that this was not necessary, and that single-sludge systems were more economical. As a result, two-sludge processes are seldom used now.

In recent years, there has been a re-emergence of a specific variation of the two-sludge process, called the *A-B process*. In the A-B process, the first stage ("A-stage") is designed to be an extremely high-rate process (~30 to 60 minutes detention time), in which a significant fraction of the influent BOD (i.e., carbon) can be removed by absorption and directed to anaerobic digestion for energy recovery. The second, or "B-stage", operates as a low-loaded, conventional activated sludge process. Because the A-stage removes 60 to 70% of the influent carbon, the volume of the second-stage reactors is small, sludge production and aeration demand are less, and the nitrification rate is higher (Versprille et al., 1984). This process configuration was invented in Europe as a method for buffering treatment facility biomasses from the harmful effects of industrial wastewaters. As originally conceived, the A-stage removes inhibitory compounds, thereby protecting the biomass in the second stage. Today, with much improved pretreatment programs, the A-B process is not needed for its original purpose.

3.2.2.2 Attached-Growth Processes

Nitrification can be accomplished in attached-growth systems such as biotowers or trickling filters, where the wastewater is circulated through a fixed media, typically plastic sheets or rocks. In these cases, about 30 to 50% of the BOD is typically removed by primary settlers located upstream of the attached-growth process. Free-floating media systems, such as moving bed biofilm reactors (MBBR), are a more modern attached-growth process that is also used for nitrification. With this type of system, which is essentially a submerged trickling filter, the bulk of the biomass is developed within the media. Figure 5.4 contains process flow diagrams for several attached-growth processes.

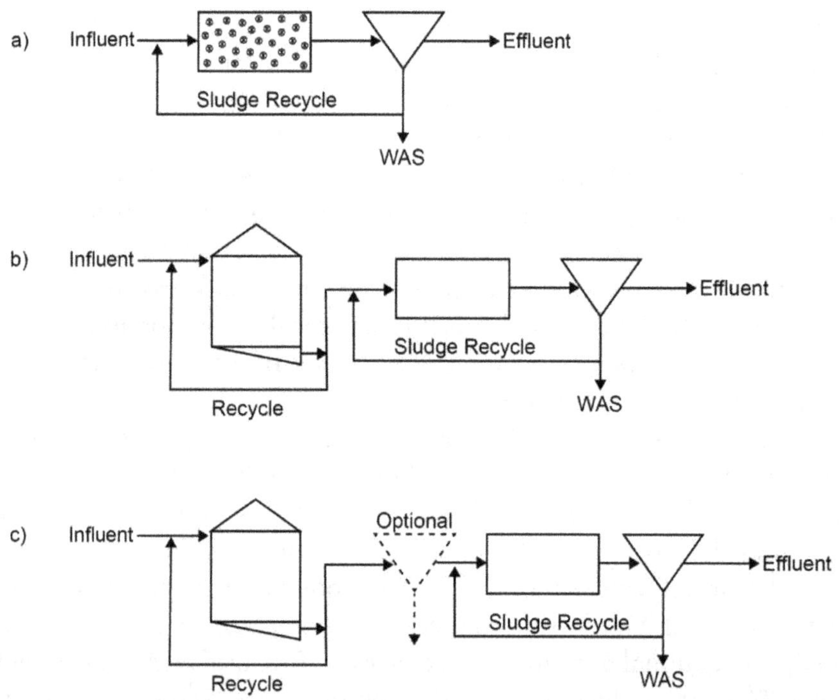

FIGURE 5.4 Hybrid processes: (a) integrated fixed-film activated sludge (IFAS), (b) trickling filter–solids contact, and (c) trickling filter–activated sludge.

3.2.2.3 Hybrid Processes

Hybrid processes grow both suspended and attached biomass in the same bioreactor. The best example is the IFAS process. In this system, free-floating or fixed, structured media are submerged in the bioreactor. The systems typically differ by the type of media used. A few examples include sponges, plastic bits of various shapes and sizes, ropes, nets, and submerged plastic trickling filter media.

3.2.3 Denitrification

Denitrification processes include anoxic bioreactors of varying configurations in combination with aerobic bioreactors. The names used for anoxic (denitrification) bioreactors indicate their position in the process relative to the main aerobic (nitrification) bioreactor. Anoxic bioreactors located upstream of the aerobic bioreactors are termed *preanoxic* or *first anoxic bioreactors*, while those located downstream are called *postanoxic* or *second anoxic bioreactors*. The two-stage anoxic–aerobic BNR process is commonly known as the *modified Ludzack–Ettinger process* (MLE), while the four-stage BNR process (anoxic–aerobic–anoxic–aerobic) is known as the *Bardenpho process*. Anoxic bioreactors located after the activated sludge process are called *tertiary bioreactors*. Conceptual process flow diagrams for suspended growth processes for nutrient removal are provided in Figure 5.5, and, for tertiary nutrient removal processes, in Figure 5.6.

3.2.3.1 Suspended Growth Processes

Most suspended growth BNR processes for nitrogen removal use a preanoxic configuration to accomplish denitrification. Because ammonia is transformed to nitrate in the aerobic reactors, with preanoxic configurations the biomass in the aerobic bioreactors (mixed liquor) must be pumped to the preanoxic zones to bring the nitrate into contact with the carbon source (BOD in the raw wastewater or primary clarifier effluent). Up to about 60% nitrogen removal can be achieved with BNR processes using preanoxic reactors; however, to obtain this efficiency, nitrate return (mixed liquor recycle) flowrates up to 4 times the influent flow are required.

The carbon source for denitrification is mostly provided by the influent wastewater. With BNR processes using preanoxic zones, the addition of supplemental carbon is seldom needed unless the ratio of BOD to nitrogen in the reactor influent is unusually low. This is a significant advantage of this process. Other advantages include the ability to recover alkalinity and slightly reduced aeration requirements. Some disadvantages include the energy costs associated with mixed liquor recirculation (MLR) pumping and

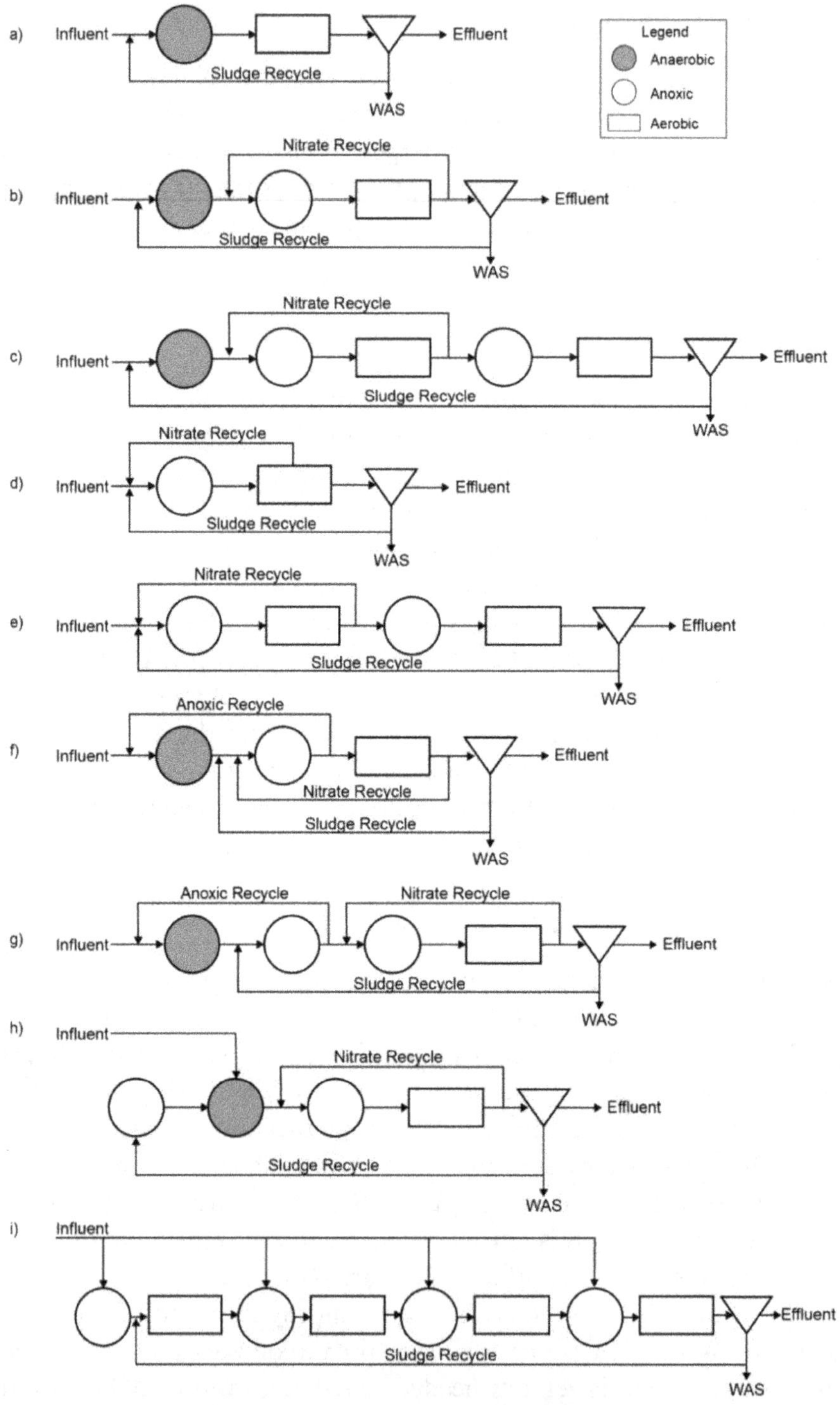

FIGURE 5.5 Suspended growth processes for biological nutrient removal (BNR): (a) anaerobic–oxic (A/O) (Phoredox), (b) anaerobic–anoxic–oxic (A2/O) (c) modified Bardenpho (B5), (d) modified Ludzack-Ettinger (MLE), (e) Bardenpho (B4), (f) University of Cape Town (UCT), (g) modified University of Cape Town (UCT), (h) Johannesburg process (JHB) (Block & Hong), and (i) step-feed biological nutrient removal (BNR) (courtesy of NACWA © 2010) (WAS = waste activated sludge).

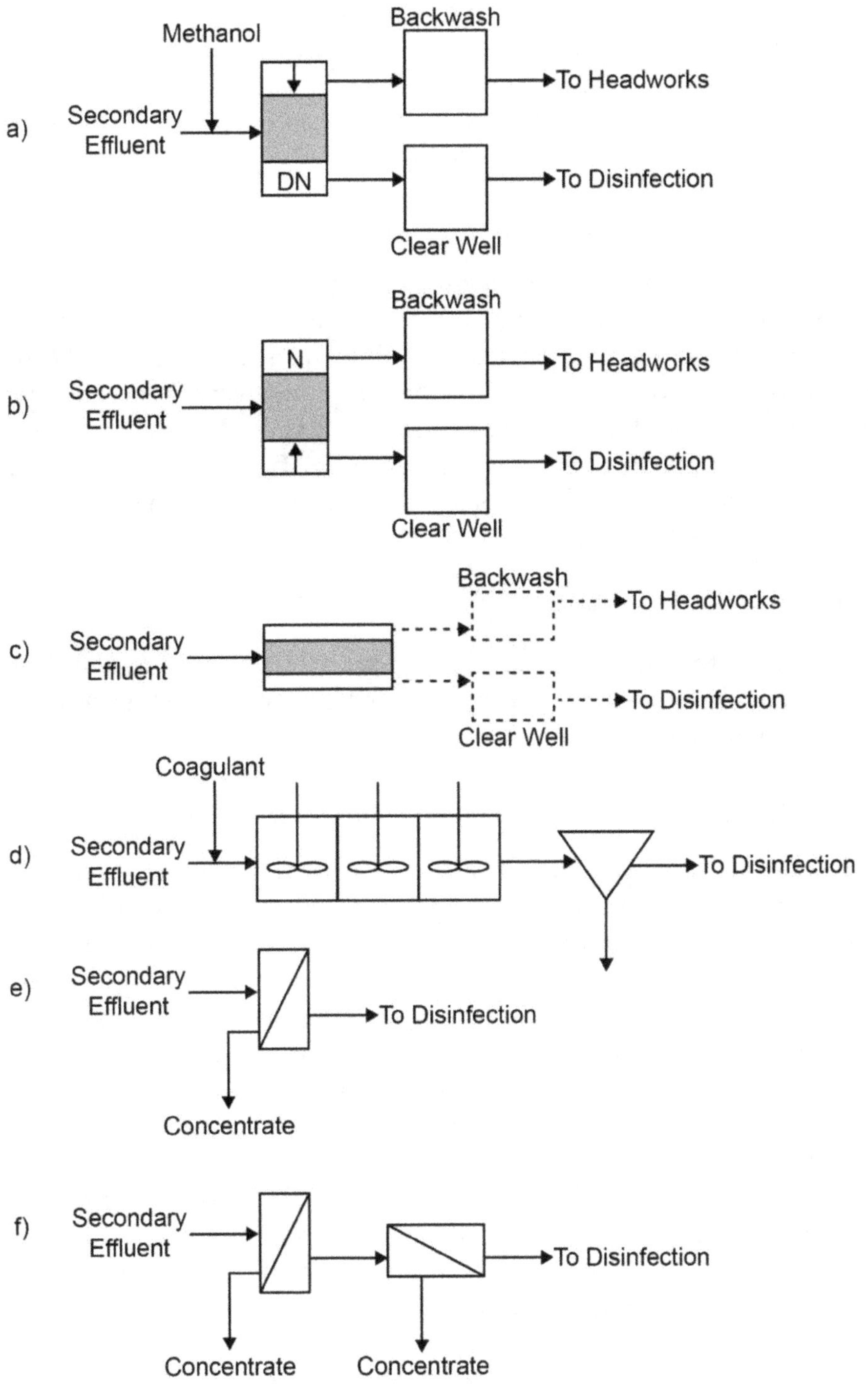

FIGURE 5.6 Tertiary treatment processes for nutrient removal: (a) denitrification filter, (b) nitrification filter, (c) filtration, (d) chemical clarification, (e) microfiltration or ultrafiltration, and (f) microfiltration (ultrafiltration)/reverse osmosis.

the inhibition of denitrification if excessive dissolved oxygen is recycled from the aerobic zone to the anoxic zone in the MLR with the nitrate.

The addition of a conventional postanoxic denitrification stage to the process train can produce effluent total nitrogen concentrations less than 3 mg/L N (Tchobanoglous et al., 2014).

3.2.3.2 Attached-Growth Processes

Attached-growth processes for denitrification include biologically active filters, MBBRs, and fluidized bed reactors. In most cases, attached-growth bioreactors for denitrification are located downstream of the nitrification system as stand-alone processes. In systems like these, carbon must be added for denitrification. Free-floating media systems like MBBRs can also be designed with anoxic zones for denitrification. Typically, these zones are placed before and after the aerobic zone in a similar manner as that for suspended growth processes.

3.2.3.3 Hybrid Processes

Similar to both suspended and attached-growth processes, hybrid (IFAS) systems can be designed to incorporate anoxic zones, which may or may not contain media.

3.2.4 Simultaneous Nitrification and Denitrification

Simultaneous nitrification and denitrification is the name given to the BNR process when both nitrification and denitrification are accomplished in the same bioreactor by operating the "aerobic" bioreactor at low dissolved oxygen concentrations. Simultaneous nitrification and denitrification processes typically operate at dissolved oxygen concentrations between 0.3 to 0.7 mg/L (Jimenez et al., 2010). Achieving and designing effective SND can be challenging. However, this type of system provides advantages such as the need for only a single type of reactor, no mixed liquor return, a reduced need for carbon, lower aeration requirements, and a better alkalinity balance. Simultaneous nitrification and denitrification systems are typically simpler to construct and, in many circumstances, can be easier to operate. On the other hand, SND processes typically require larger bioreactors and may require more sophisticated sensors and aeration system controls.

3.2.5 Shortcut Nitrogen Removal (Nitritation–Denitritation) and Deammonification

A newer approach to nitrogen removal developed over the last two decades relies on "shortcut" pathways in the nitrogen cycle (nitritation/denitritation and deammonification) to accomplish nitrogen removal. These pathways are only achievable using tight control of the treatment process under specific environmental conditions. Nevertheless, processes based on shortcut pathways are gaining popularity, particularly for treatment of sidestreams from dewatering anaerobically digested sludge, because they require less aeration and less carbon addition than conventional nitrogen removal technologies. Some examples of shortcut nitrogen removal processes include the Sharon™, Anammox™, combined Sharon™/Anammox™, Demon™, and ANITA™ Mox processes.

3.4 Phosphorus Removal

Phosphorus removal is accomplished by chemical means, biological means, or a combination of biological and chemical means.

3.4.1 Chemical Phosphorus Removal

Chemical technologies for phosphorus removal rely on coagulation and precipitation of the phosphorus as a sparingly soluble compound through the addition of metal salts (i.e., iron salts and aluminum salts) or lime. Chemical phosphorus removal can be implemented at different locations in the treatment train. The common terminology for these locations is *preprecipitation*, *simultaneous precipitation*, and *postprecipitation*. With preprecipitation, chemicals are added to primary treatment. This is also referred to as *chemically enhanced primary treatment*. With simultaneous precipitation, the chemicals are added to the biological process. *Postprecipitation* refers to addition of the chemicals to the mainstream after the biological treatment process (tertiary chemical addition). The advantages and disadvantages of the various chemicals and locations for adding them are well documented elsewhere (Srivastava and Sanghi, 2005).

3.4.2 Biological Phosphorus Removal

Phosphorus is removed from wastewater by incorporation to the biomass and wasting of the phosphorus-laden excess biomass (i.e., waste activated sludge) from the treatment process.

Biological phosphorus removal relies on special groups of bacteria that are capable of taking up phosphorus in excess of their growth needs when they are exposed to anaerobic conditions in the presence of volatile fatty acids (VFAs) followed by aerobic conditions. Volatile fatty acids are organic acids typically generated in wastewater collection systems by the decomposition of organic material under anaerobic conditions; however, treatment facilities that are deficient in VFAs can generate them on-site using a process called *fermentation*. Bacteria capable of taking up phosphorus in excess of their basic metabolic needs are commonly referred to as *polyphosphate-accumulating organisms* (PAOs). Under anaerobic conditions, PAOs take up VFAs (primarily acetic acid and propionic acid) from wastewater and store this carbon within their cells. Subsequently, under aerobic conditions, PAOs use their carbon reserves for growth, cell maintenance, and creation of phosphorus reserves within their cells. Because phosphate is then taken up from the wastewater and stored inside the cells as polyphosphates, the phosphorus content of the sludge is increased significantly. Polyphosphate-accumulating organisms with high phosphate content are separated from the clean water along with the rest of the biomass by sedimentation and ultimately removed from the system when excess biomass is removed from the process. The BNR processes for phosphorus removal typically incorporate an anaerobic zone upstream of the aerobic bioreactors to create the required anaerobic–aerobic sequence. The most established processes for biological phosphorus removal are the two-stage anaerobic–oxic (A/O) process, the three-stage anaerobic–anoxic–oxic (A2/O) process, and the five-stage (or modified) Bardenpho process.

3.4.3 Biological Phosphorus Removal with Chemical Trimming

Chemical phosphorus removal technologies are typically combined with biological phosphorus removal technologies to maximize phosphorus removal. Oftentimes, chemical phosphorus removal is used to trim effluent phosphorus concentrations from biological phosphorus removal systems to meet stringent limits. Supplemental chemical addition followed by filtration is a common practice to meet stringent phosphorus limits.

3.5 Combined Nitrogen and Phosphorus Removal

Combined biological nitrogen and phosphorus removal is accomplished by technologies developed to provide both nitrogen and phosphorus removal in one system. This typically involves incorporation of three stages as a minimum: anaerobic, anoxic, and aerobic. Technologies developed for biological nitrogen and phosphorus removal include the A2/O process, the University of Cape Town (UCT) processes, the Johannesburg (JHB) process, and the modified (five-stage) Bardenpho process. Detailed information on the various BNR processes can be found in literature by Grady et al. (2011), Droste (2009), Henze et al. (2008), Henze et al. (2002), Tchobanoglous et al. (2014), van Haandel and van der Lubbe (2012), WEF et al. (2009), and WEF (2010).

4.0 THE FOUNDATION—SET OBJECTIVES (FOR THE LOCAL AUTHORITY)

The most successful endeavors are generally those that begin with a solid understanding of the desired final objectives. The same is true when selecting a technology to remove nutrients. In addition to deciding which nutrients must be removed and to what concentrations, policy decisions should be made as to the key technical, environmental, social, and economic characteristics that the final facility must possess.

4.1 Regulatory Driven

As discussed in more detail in Chapters 2 and 3, wastewater treatment goals are strongly driven by state and federal environmental laws and regulations. It is often said that owners can be certain of two things about water quality regulations: (1) they will change and (2) they almost always become more stringent with each passing permit cycle. As a result of U.S. EPA's efforts to have states implement numeric standards for maximum instream nutrient concentrations (also known as *numeric nutrient criteria* or *numeric nutrient endpoints*), many more WRRFs will be given limits on their discharge of nutrients over the next decade. Often, the regulatory agencies not only decide which nutrients to remove and to what concentrations, they also dictate the timetable for performance. Perhaps the most difficult situations, however, are those where nutrient removal is expected in the future, but the specific criteria and timetable have not been established. Owners of facilities about to undergo expansion must then decide whether to implement nutrient removal during the expansion, and to what standard, or wait.

Nutrient removal technologies are flexible, and, with proper planning, a facility can be designed and built so that it can be upgraded when required. To be prepared for the future, however, facility owners must understand local water quality challenges and the specific agencies with jurisdiction in their area and plan for likely future changes in regulations.

4.2 Considerations in Technology Selection

Although many facilities face similar challenges, it is nearly impossible to find one nutrient removal technology that is universally best. The selection of the specific needs that must be met by a nutrient removal technology is a policy decision to be made by facility owners. Although economic factors are typically the principal consideration, other criteria are often of overriding importance. For example, at existing facilities in urban areas, the land available for new facilities is often at a minimum. In such situations, the amount of land required for a BNR technology can become a principal concern. Other possible considerations may include the type of wastewater collection system (combined or separate), seasonal loads from tourists or agricultural operations, and the quantity and type of industrial wastes that are sent to the facility. A combined collection system coupled with nutrient limitations can be an expensive combination unless occasional storm bypasses, potentially with physical/chemical treatment, can be permitted. Heavy industrial contributions may force a facility to adopt more expensive physical or chemical technologies unless substances toxic to nitrifying bacteria can be effectively controlled.

In areas of the country where reuse is a viable option, additional forces may drive the selection of treatment methods. For example, many reuse systems do not have the demand to use the entire flow from the treatment facility. This is particularly true when reclaimed water is used for irrigation where seasonal variations in rainfall, temperature, and crop growth will often force short-term discharges to surface water. Wet or cold weather storage can be a possible alternative where sufficient land is available to store water for the expected duration of inclement conditions. Aquifer storage and recovery is another alternative where geologic conditions permit. Aquifer storage and recovery, when allowed by local authorities and depending on the aquifer, may require additional treatment beyond secondary treatment (for example 10 mg/L nitrate, filtration, or high-level disinfection). Otherwise, sufficient treatment capability must be provided to meet the generally more stringent criteria for surface water discharge. The health criteria associated with public access irrigation coupled with the need for a seasonal surface water discharge can lead a utility owner to a different BNR process than might otherwise be chosen.

5.0 THE FIRST STEP—ESTABLISH BASIC REQUIREMENTS

The first step in selecting a nutrient removal process is to establish basic requirements. These requirements include effluent criteria, influent characteristics, the methodology to be used to compare processes, and the selection criteria to be used to rank and compare technologies.

5.1 Define Effluent Requirements

The water quality criteria set forth in a facility's operating and discharge permits are perhaps the most important factors to be used in selecting potential treatment processes. Although it is sometimes necessary to begin down this path without explicit permit limits, it is better if effluent limits are known before embarking on a nutrient removal project to minimize the risk of unplanned expenditures. Much can change with respect to the interpretation of verbal or vaguely written regulatory requirements by owners, project designers, and regulatory personnel during the 3 to 5 years or more that it takes to implement a facility construction project. Some typical nutrient limits currently in place in the United States are given in Table 5.3, while typical levels or tiers for numeric criteria for nitrogen and phosphorus are summarized in Table 5.4.

TABLE 5.3 Typical nutrient limits.

	Criterion (as equivalent annual average or monthly average concentration)								
Location	Total ammonia	Un-ionized ammonia	10 to 15 mg/L nitrate as N	5 to 8 mg/L total nitrogen	2 to 3 mg/L total nitrogen	< 2 mg/L total nitrogen	1 to 2 mg/L total phosphorus	< 0.5 mg/L total phosphorus	< 0.2 mg/L total phosphorus
Chesapeake Bay	X				X			X	
Great Lakes	X						X	X	
Florida			X		X		X		
Puget Sound (Budd Inlet)					X*				
Colorado (Regulation 85)	X		X*	X*			X		
Spokane River	X								X
Chattahoochee River	X							X	X
New England	X			X			X	X	X
Kansas	X			X			X		
Texas	X						X	X	
Idaho (Lower Boise and Snake Rivers)									X
Southern California (Santa Ana River)	X		X*						
San Diego Region				X		X			X

*Total inorganic nitrogen–ammonia plus nitrate plus nitrite

TABLE 5.4 General structure of common tiers for nutrient criteria. (BNR = biological nutrient removal; A/O = anaerobic–oxic; MLE = modified Ludzack-Ettinger; A2/O = anaerobic–anoxic–oxic; JHB = Johannesburg process; SND = simultaneous nitrification and denitrification; AOP = advanced oxidation process)

Nutrient criteria tier	Common nitrogen criteria (mg/L)	Common phosphorus criteria (mg/L)	Common technologies to achieve tier limits
Tier 1	$NH_3 \leq 1$ to 3	N/A	Requires complete nitrification unless treatment split between multiple parallel trains and effluents combined.
Tier 2	Total nitrogen ≤ 10 to 12	Total phosphorus ≤ 1	Requires complete nitrification and denitrification to 7 to 9 mg/L of NO_3. Phosphorus limits can be met biologically or with chemicals. Effluent filtration typically not required. Tier 2 limits can be met biologically using BNR processes with two- or three-stage anaerobic/anoxic/aerobic processes such as A/O, MLE, A2/O, and JHB.
Tier 3	Total nitrogen ≤ 6 to 8	Total phosphorus ≤ 0.3 to 0.5	Nearing minimum limits that can be met with basic two- and three-stage BNR processes. Meeting total nitrogen limit may require anoxic zones or reactors after the main aeration zone or SND. Effluent filtration is required for reliable removal of phosphorus in effluent solids.
Tier 4	Total nitrogen < 3	Total phosphorus < 0.3	Close to minimum limits with current technologies. Meeting total nitrogen limit requires anoxic zone after main aeration zone or tertiary nitrogen removal. May require supplemental carbon addition. Typically, biological P removal with chemical P removal and filtration are required to reach total P values.
Tier 5	Total nitrogen ≤ 2	Total phosphorus ≤ 0.1	U.S. EPA's numeric nutrient criteria for maximum instream concentrations will likely become end-of-pipe limits in some watersheds. Complex and costly technologies may be required for stringent limits. To achieve ultra-low nitrogen and phosphorus limits, advanced technologies like membranes or AOP may be required.

The performance of nutrient removal treatment technologies is highly dependent on the characteristics of the wastewater being treated, much more so than conventional activated sludge treatment. It is, therefore, prudent to establish the characteristics of the wastewater to a higher degree than one might ordinarily do for the design of a secondary treatment facility. This is particularly important when attempting to produce an effluent with concentrations of total nitrogen at or below 3 mg/L, or concentrations of total phosphorus below 0.5 mg/L.

The minimum wastewater temperature that will be experienced at the facility is a critical factor in the design of a BNR process. Thus, particular attention should be paid to the need for seasonal or year-round nutrient removal. Owners should consider negotiating for seasonal permits when it is clear that winter discharges will have minimal effect on the receiving water. The duration and calendar location of nutrient removal seasons should also be given close scrutiny. It saves little money to design for nitrification or nitrogen removal beginning on the first of March or April because the nutrient removal process would need to start up during the coldest water temperatures of the year, which typically occurs February and the beginning of March.

5.2 Characterize Influent

It is common for municipal wastewater facilities to monitor only for the specific water quality parameters listed in their permits. However, a facility should begin regular monitoring for nitrogen and phosphorus compounds whenever it appears likely that nutrient limits may be imposed in the future. This will ensure that if or when the time comes to implement nutrient control, the facility designer will have full knowledge of the mass loadings to be treated. Many owners and operators are hesitant to do this for fear of causing nutrient limits to be enacted simply because they are reporting the information, or in a desire to keep laboratory costs to a minimum. The former is generally faulty reasoning, as adverse water quality effects on the receiving waters will ultimately become known to the permit writers regardless of a lack of monitoring on the part of the owner. Any savings on the costs of monitoring will be completely overwhelmed by the conservatism applied by the design engineer if he or she does not have adequate knowledge of the nutrient loadings to the proposed facility.

5.2.1 Flows/Loads

Diurnal flow variations can pose a potential concern for enhanced biological phosphorus removal (EBPR) if such variations also result in low influent concentrations of readily biodegradable organic carbon compounds (readily biodegradable chemical oxygen demand [rbCOD]) and, specifically, VFAs. However, such variations may also be associated with lower phosphorus concentrations, in which case process stability may not be affected. Implementation of raw wastewater or primary solids fermentation can stabilize the readily available carbon-to-phosphorus ratios and help maintain EBPR stability. Alternately, external carbon sources (e.g., crude glycerol [Coats et al., 2015]) can be considered to stabilize the process.

5.2.2 Carbon Types, Magnitude, and Ratios

Denitrification requires organic carbon as the electron donor; the form of organic carbon is not critical, and can generally be measured by gross parameters such as BOD or chemical oxygen demand (COD). Conversely, successful EBPR requires sufficient VFAs to drive the necessary reactions such that bacteria will accumulate excess phosphorus. Moreover, propionate is suggested to be more favorable for EBPR (Oehmen et al., 2007), yielding mixed liquor suspended solids more enriched for PAOs over glycogen accumulating organisms (GAOs). Recent research suggests that crude glycerol can also be used in lieu of VFAs (Coats et al., 2015).

The ratio of influent VFA-to-influent phosphorus ($VFA_{infl}:P_{infl}$) has been suggested as a potential indicator for achieving successful EBPR (Grady et al., 2011), with minimum ratios ranging from 15:1 on a VFA (as COD):P basis (Horgan et al., 2010) to 18:1 on a rbCOD:P basis (Tchobanoglous et al., 2014). As VFA:rbCOD increases, the required rbCOD:P to maintain EBPR stability decreases.

5.2.3 Nutrient Loads

Influent wastewater should be characterized for all nitrogen and phosphorus species, both present in solution and as particulate matter. Sidestreams should be characterized similarly. All nutrient loads should be established on both a concentration and mass basis, factoring in variability in sidestream loads that might occur as a result of the daily and weekly dewatering schedule.

5.2.4 Alkalinity and pH

Nitrification both consumes alkalinity and is sensitive to pH; the pH should be maintained between 6.5 and 8, with a minimum effluent alkalinity of 50 to 100 mg/L (as $CaCO_3$) recommended (WEF et al., 2005). Denitrification, operated in a preanoxic mode, will recover approximately 50% of the alkalinity lost via nitrification, and can enhance pH stability. The pH values greater than 7.5 are understood to improve EBPR by selecting for PAOs over GAOs (Tchobanoglous et al., 2014).

5.2.5 Potential Toxicity

Ammonia-oxidizing bacteria are the most sensitive bacteria associated with nutrient removal, and can potentially serve as an indicator of toxic compounds in the wastewater. Potentially toxic compounds have been identified by U.S. EPA (2010b) and WEF et al. (2005). Excess nitrate in the return activated sludge stream that is discharged into the anaerobic zone can upset the EBPR process if insufficient organic carbon (principally VFAs) is available to reduce the nitrate and induce the required anaerobic zone reactions.

5.2.6 Temperature

In general, biological reaction rates decrease as the temperature decreases, reducing treatment efficiency (U.S. EPA, 2010b). Cold-water temperatures are a primary concern for BNR processes. While lower temperatures also adversely affect EBPR kinetics, PAOs have been shown to outcompete GAOs at temperatures below 20°C (U.S. EPA, 2010b; Tchobanoglous et al., 2014). Typically, for both nitrogen and phosphorus removal, increased solids retention time is used to maintain process stability as the water temperature decreases.

5.2.7 Wet Weather Characteristics

Influent characteristics of most concern are listed in Table 5.5, and many are discussed briefly herein. It is necessary that the design flows and loadings be developed from either master planning or historical records and trends. Good master planning will establish the contributions to the facility flow from residential, commercial, and industrial customers and infiltration and inflow. Historical analysis will develop frequency of exceedance curves for flow and mass loadings of the parameters of concern. It is desirable to analyze sufficient historical data to identify long-term trends. Typically, 3 to 5 years of data are desirable in high-growth areas, although 1 to 2 years may be adequate in more stable regions. In addition to the design flow and loads, it is essential to know the likely range of these parameters from minimum loadings at startup to sustain peak flows at design conditions. For nitrogen removal projects, it is necessary to define the total Kjeldahl loading in the influent because most of the organic nitrogen will be converted to ammonia in the treatment process. It is also essential to establish the biodegradable organics concentrations as measured by COD (5-day BOD [BOD_5]), soluble COD (BOD_5), filtered and flocculated COD, and VFAs because the organic compounds provide the bacterial food that drives the denitrification and phosphorus removal reactions in most BNR processes. If the quantity and type of organics available are out of proportion with the quantity of nitrogen to be removed, they will influence process selection and determine the need to add an external carbon source such as methanol or glycerol. For biological phosphorus removal, it is desirable to establish the range of VFAs available in the wastewater to be treated because the performance of biological phosphorus removal processes is a function of acetic and propionic acid concentrations. Biological nutrient removal treatment processes, especially

TABLE 5.5 Influent parameters desirable for design of nutrient removal facilities.

Parameter	Specific measurement	Abbreviation
Flow variability	Average day maximum month (or 90th percentile)	Q_{adf}
	Maximum day (or 99th percentile)	Q_{max}
	Peak hour (or 99.9th percentile)	Q_{peak}
Mass loading variability	Average day maximum month (or 90th percentile)	Q_{adf}
	Maximum day (or 99th percentile)	Q_{max}
Organic content	5-day biochemical oxygen demand	BOD_5
	Ultimate biochemical oxygen demand	UBOD
	Chemical oxygen demand	COD
Organic Acids	Volatile fatty acids	VFAs
Phosphorus	Orthophosphorus	OP
	Total phosphorus	TP
Nitrogen	Ammonia	NH3
	Total Kjeldahl nitrogen	TKN
	Nitrite	NO_2
	Nitrate	NO_3
Alkalinity		
pH		
Toxics	Heavy metals, organic solvents, chlorinated hydrocarbons, etc.	
Sulfur	Hydrogen sulfide	H_2S
	Sulfates	SO_4
Temperature	Minimum/maximum water temperature	
	Minimum/maximum air temperature	

nitrification, are susceptible to inhibition by low pH, toxic materials such as heavy metals, and various organic compounds such as pesticides and industrial solvents. It is essential for a BNR process to operate without frequent upsets. It is, therefore, desirable to eliminate the addition of toxic materials to the collection system through the development and enforcement of an industrial pretreatment program.

5.3 Establish Evaluation Methodology

There are many evaluation and selection methods available to assist in selecting a treatment technology for implementing nutrient removal. Ideally, one would thoroughly examine in detail all possible technologies, including variations and combinations; however, limitations on time and budget will always preclude this level of effort. Still, there are a few general concepts that should be present in whichever methodology is chosen. The main thrust for any selection method is to make one "best" choice within the budget and time available, and without missing desirable possibilities or key considerations. The magnitude of this effort is defined by the owners, and should be sufficient to provide desired comfort that the right choices are being made. The scope of the evaluation should consider factors such as the overall cost of the decision, the effect of the decision on the utility or community's long-term plans, engagement of the community in the decisions, and the likelihood of disagreement with the decision. In general, as the effect of a decision increases, the level of effort for the evaluation should also increase.

The best evaluation methods are rational, objective, and scientific. Statistics and mathematics, however, should never be allowed to replace human values. The ultimate decision on the selection of a nutrient removal technology must be firmly founded in those features judged most important to the community served by the facility.

It has been proposed by some environmental planners (McAllister, 1982) that selection procedures include two main groups of activities. The first is called the *analysis phase*, wherein the technical effects of each technology are identified and evaluated. The second phase is the synthesis phase, wherein all of the effects are integrated to a single viewpoint or rank representing each technology or process. This concept is illustrated in Figure 5.7. The first phase is objective, technical, and quantitative, while the second is subjective, holistic, and qualitative.

5.3.1 Arbitrary Methods

Commonly used evaluation methods can be placed in three main groups. These are arbitrary, informal, and formal methods, which are each described in more detail in this section and Sections 5.3.2 and 5.3.3.

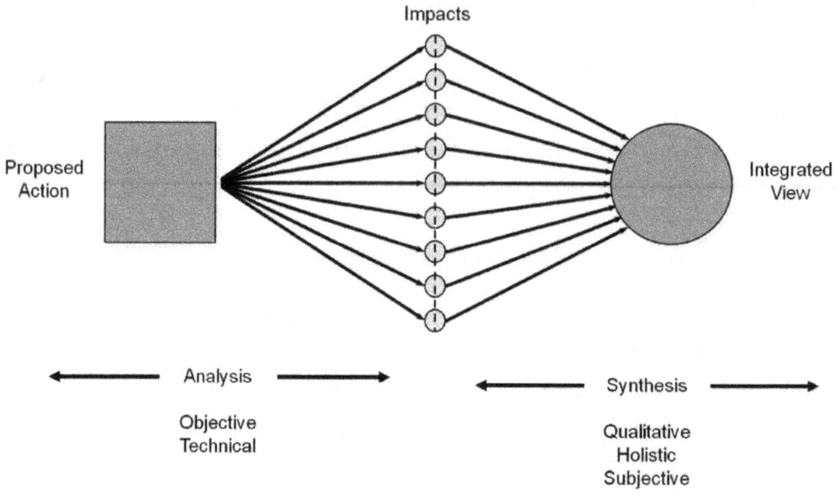

FIGURE 5.7 Analysis and synthesis phases of alternatives evaluations.

Arbitrary methods can be distinguished by a lack of any objective rationale or criteria. Because of the lack of structure or guidelines with arbitrary methods, different evaluators can come to completely different decisions. As a result, evaluators can be easily swayed, or can intentionally sway, results in favor of their own values, biases, and agendas. In the end, the owner cannot know if a technologically appropriate and economically desirable selection was made. For these reasons, technology selections based solely on vendor recommendations, regional practices, or recommendations made during procurement of professional services should be avoided.

5.3.2 Informal Methods

Informal selection methods are those based on logic and reason; however, they lack the structure of formal methods. Informal methods rely heavily on personal review and reflection. For many people, this will include discussion and debate between experts and interested individuals. Informal methods are certainly valid selection methods; however, they tend to be imprecise and subjective by nature.

5.3.3 Formal Methods

Today's demands for comprehensive, legally acceptable answers, and at least the appearance of extreme accuracy, lead many treatment facility owners to undertake one of the formal, published evaluation methodologies. Many of the formal methods are more comprehensive and substantive. The principal formal evaluation methods are listed in Table 5.6. Several of these methods

TABLE 5.6 Formal evaluation methods.
Cost–benefit analysis
Professional (expert) judgment
Environmental evaluation system
Judgment effect matrix
Planning balance sheet
Goals achievement matrix
Energy analysis
Delphi method
Triple bottom line plus (TBL+)
Life cycle analysis
Ecological footprint

were developed specifically for the evaluation of large environmental and public works projects. The U.S. Army Corp of Engineers uses the cost–benefit analysis method routinely, and the judgmental impact matrix method was developed especially for a regional wastewater treatment and disposal project in the Chicago, Illinois, area.

The formal methods are typically characterized by elaborate and detailed procedures for determining the environmental, social, economic, and political effects of each alternative. The summation of these effects then serves as the basis for the evaluation. This summation can be accomplished using a number of approaches. For instance, a ranking system that assigns numerical scores to each alternative for specific criteria and then weights these criteria based on their relative importance to the project can be used as part of a formal evaluation. While this is a valid and reasonable approach, it is common for there to be disagreements on the relative weighting of factors and, therefore, the overall outcome of the evaluation. A more robust approach to formal evaluation is to equate all criteria to the same equivalent unit, which is often dollars. This allows an "apples-to-apples" comparison of each alternative.

5.4 Establish Selection Criteria

Perhaps the most important part of a formal evaluation is the determination of the selection criteria, against which any particular technology or action will be evaluated. While varied, selection criteria typically can be categorized in one of four areas:

- Technology,
- Finance,
- Environment, and
- Social.

The exact criteria selected in each category will vary by the circumstances and priorities for each utility. For example, in an urban environment, the noise or odor effects of an alternative may be an important criterion under environmental effects. This criterion may not be important in environments that are more rural. While the selected criteria will vary by utility, it is always important that the selected criteria align with the utility's goals. This alignment provides additional support for the evaluation and the selected outcome.

6.0 THE NEXT LEVEL—CONCEPTUAL PROCESS COMPARISONS

6.1 The Universe of Possibilities

There is much to choose from in the world of nutrient removal. Possibilities range from tried and true, on-site, concrete and steel suspended growth processes such as the Bardenpho and A2/O processes, to more innovative approaches using submerged packed-bed reactors or off-site aquatic- or land-based systems. At the beginning of the selection process, it is helpful to generate a "long list" of possible technologies that could be implemented to provide a specified degree of nutrient removal at a wastewater facility. To provide nitrification and nitrogen removal, the universe of possibilities is large, particularly when one considers combinations of removal processes and integrated approaches. For phosphorus removal, the list is not quite as big, but still large enough. Figures 5.8, 5.9, and 5.10 illustrate many of the currently available nutrient processes, including chemical and physical treatment processes. These figures are in the form of taxonomic charts or graphical trees that group available processes based on logical relationships. The main groupings include

- Suspended growth processes,
- Fixed-growth processes,
- Hybrid processes,

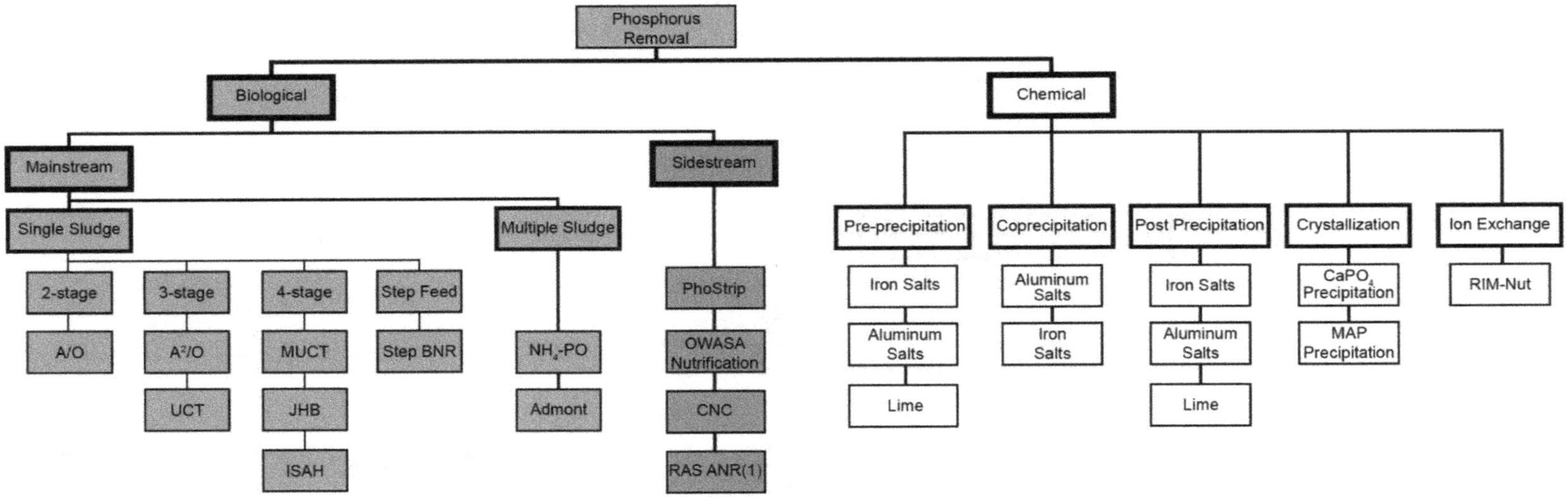

FIGURE 5.8 Phosphorus removal technology tree (BNR = biological nutrient removal; A/O = anaerobic–oxic; A2/O = anaerobic–anoxic–oxic; MUCT = modified University of Cape Town; UCT = University of Cape Town; JHB = Johannesburg process; CNC = Charlotte North Carolina; OWASA = Orange Water and Sewer Authority; MAP = magnesium-ammonium-phosphate; RAS = return activated sludge).

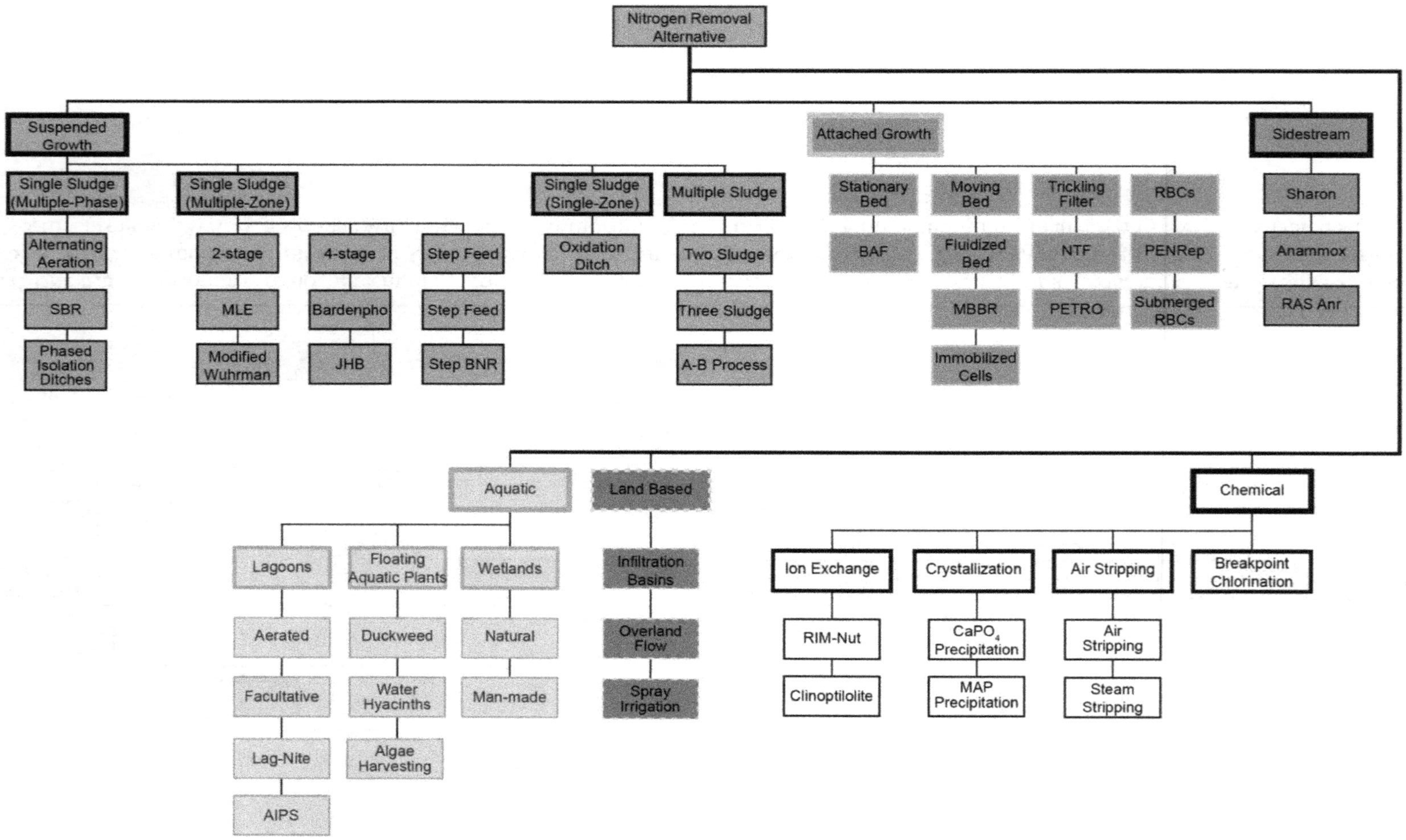

FIGURE 5.9 Nitrogen removal technology tree (SBR = sequencing batch reactor; MLE = modified Ludzack-Ettinger; JHB = Johannesburg process; AIPS = advanced integrated pond systems; BAF = biological aerated filter; NTF = nitrifying filters; RBC = rotating biological contactor; RAS = return activated sludge).

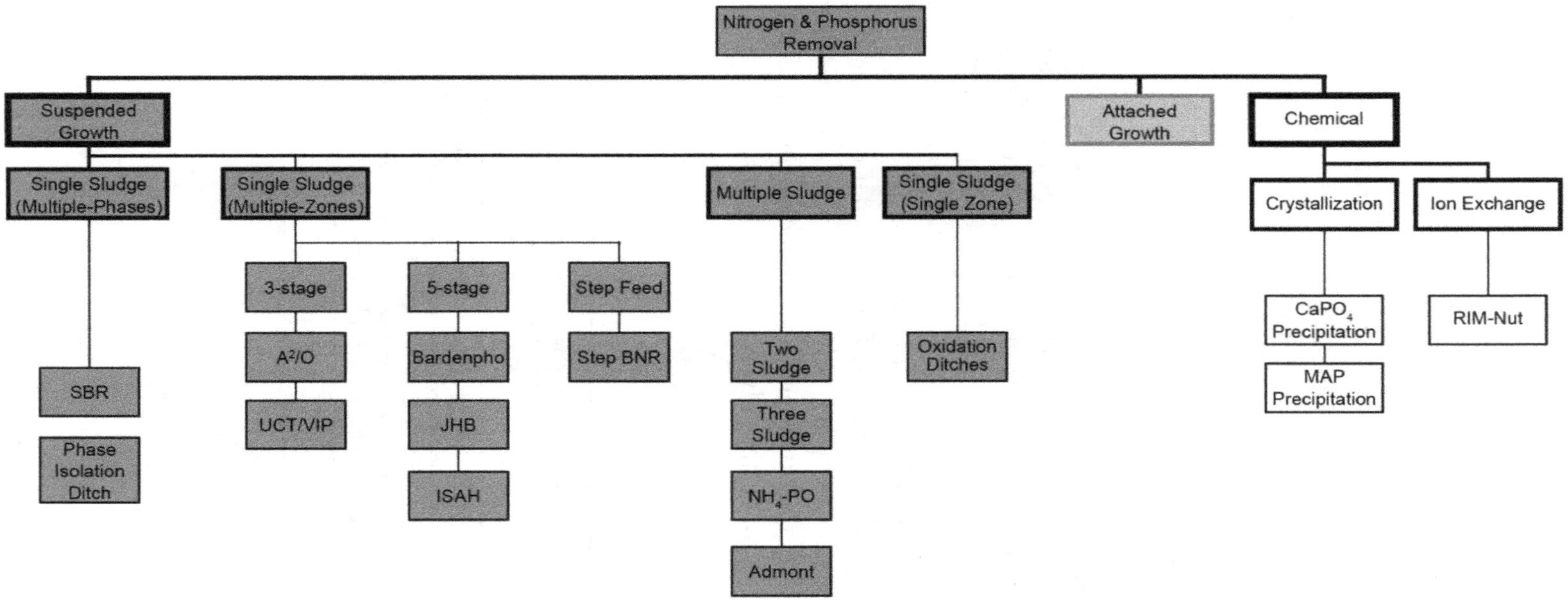

FIGURE 5.10 Technology tree for nitrogen and phosphorus removal (SBR = sequencing batch reactor; A2/O = anaerobic–anoxic–oxic; BNR = biological nutrient removal; UCT = University of Cape Town; VIP = Virginia initiative plant; JHB = Johannesburg process; BNR = biological nutrient removal; MAP = magnesium-ammonium-phosphate; RAS = return activated sludge).

- Land-based methods, and
- Aquatic-based methods.

The first two groupings can also be described as on-site treatment methods, while the latter two groupings are typically off-site methods because they can require large areas of land that is often remotely located from the secondary portion of the treatment facility.

6.2 Screening

With the plethora of alternatives described in the previous section, a systematic screening method is often needed to reach a short list of alternatives for more thorough evaluation. The formal methods of evaluation will often use one or more successive levels of screening to quickly narrow the list of possibilities to a manageable number to be evaluated in greater depth. Figure 5.11 illustrates this concept. In evaluation methods based on expert judgment, the final short list can sometimes be reached quickly depending on the selection criteria established. This method is often seen in the reports of consulting engineers who will select three or four processes for detailed evaluation based on their own professional judgment of a client's situation. One should be cautioned about being too quick to arrive at a short list. The

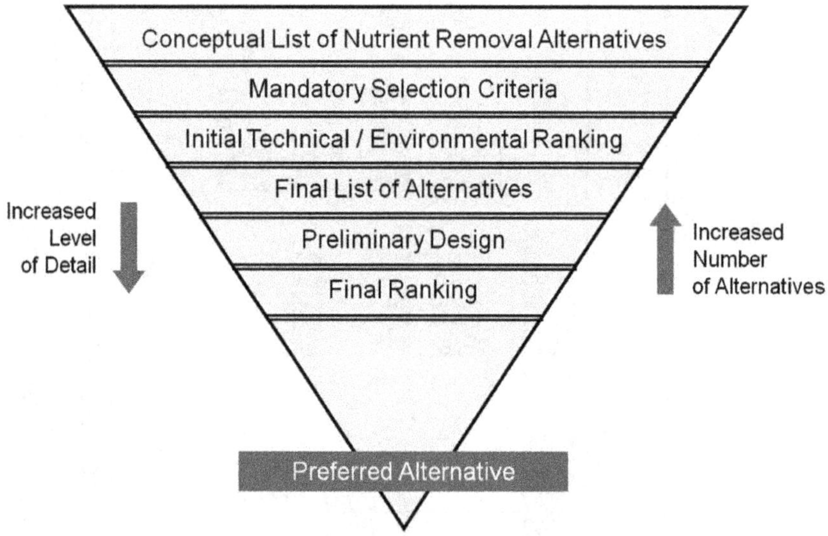

FIGURE 5.11 Example screening procedure used to narrow the field of nutrient removal technologies.

additional time required to work through a long list need not be great and could pay exceptional dividends.

Useful filters that can be used to screen a long list include the ability of a process to meet the specified effluent quality, the acceptability of the process to the regulatory agency, and the existence of full-scale facilities successfully using the process. Subsequent levels in a selection process can include preliminary technical and environmental evaluation and screening, preliminary designs and cost estimates, value engineering, and a final evaluation and ranking of alternatives, options, or scenarios.

Evaluation criteria can be established for any subject of particular concern to a facility owner. When the purpose of a selection is to pick a treatment process to provide nutrient removal or recovery, however, there are certain subjects of concern to most locations. These are listed in Table 5.7, and are discussed briefly in this section.

The ability of a particular technology to consistently produce a treatment facility effluent with concentrations of nitrogen or phosphorus below the specified levels is paramount to any facility owner. The removal efficiency of some nutrient removal processes is directly related to the influent characteristics, especially the relative quantity and type of biodegradable organic carbon to the quantity of nitrogen and phosphorus to be removed. Thus, it is important to consider this factor when selecting a process. A process should

TABLE 5.7 Potential screening criteria for use in technology selection studies.

Performance potential
Operating experience
Operating complexity
Chemical usage
Energy requirements
Land requirements (footprint)
Sludge production
Aesthetics
Equipment types
Use of existing facilities
Annual costs
Capital cost

be promptly eliminated from consideration if there are serious doubts about its ability to meet permit limits. In some instances, it may be advisable to conduct testing on a bench, pilot, or demonstration scale to be completely certain of a process's ability to perform with a particular wastewater.

The quality and amount of experience that exists within the wastewater treatment community is another important consideration. The degree of assurance that can be given that a technology will adequately perform increases with the number of installations in service and the length of time they are in service. Everyone knows and is comfortable with the abilities of the conventional activated sludge process. It has been used for decades in all sizes of treatment facilities in just about every region in the world. While the record of experience of BNR processes does not yet compare with activated sludge, the installed experience base of conventional BNR processes is substantial.

Decades ago, physical and chemical processes were investigated by a number of researchers and communities; however, by the late 1960s and early 1970s, only chemical phosphorus removal came into common use. After the invention of the MLE and the Bardenpho processes by James Barnard in the mid-1970s, BNR quickly became the nitrogen removal technology of choice. The first Bardenpho process was introduced to the United States in 1979. Just over a decade later, there were more than 100 U.S. treatment facilities using suspended growth nutrient removal processes. The number of operating WRRFs using BNR processes has grown significantly greater since then. There are fewer treatment facilities using attached-growth processes for nutrient removal; however, they offer significant advantages in many instances.

There is always someone marketing a treatment process in this country who can point to dozens of operating facilities in a foreign country. While many significant wastewater treatment innovations have originated overseas, the designer and owner should thoroughly evaluate the experience of such technologies. The differences between successful operation overseas and success in this country are often subtle. Differences in facility size, wastewater characteristics, operating staff, and operational methods can be important. The local design customs and practices can also vary from the engineering methods used in the United States.

The operational complexity of a process can be a big concern, particularly with the owners of smaller facilities. Processes with numerous pieces of mechanical equipment and those that require quick operator response to changes in the influent quality will require more routine maintenance and operator time. Those that rely on sophisticated electronic sensors and controllers require regular maintenance of the sensors and controllers in addition to adequate backup units.

The use of lime or metal salts for phosphorus removal or the addition of supplemental carbon feed for nutrient removal may be necessary or desirable in some situations; however, such use often implies a trade-off between capital and operating costs. The facility owner should be conscious of such trade-offs. During facility upgrades, the equipment and process type may be quite different between the existing facility and a proposed upgrade. It is often undesirable, and may be unacceptable to some operational staffs, to construct a process upgrade using equipment requiring different parts and maintenance methods or requiring the operators to learn completely new operating procedures.

Sludge production and handling requirements can differ between alternative processes. For example, chemical phosphorus removal can produce significant additional quantities of chemical sludge (metal phosphates and hydroxides); however, there is little worry about returning phosphorus to the process in the sidestreams. Biological phosphorus removal processes also generate additional sludge, but special care must often be taken to be certain that the loosely bound phosphorus is not inadvertently released and returned to the process as a result of sludge treatment.

Increasingly, communities, particularly treatment facility neighbors, are concerned about the visual, olfactory, and noise effects of treatment facilities. When these factors are of special interest, evaluation criteria are best included that address these features of a nutrient removal process. Often, the aesthetic features of BNR processes are similar to those of WRRFs in general. Exceptions to this include the visual effect of a nitrifying trickling filter, concerns about odors that could be released from a sidestream fermentation tank, or noise from added mechanical equipment.

7.0 COST COMPARISONS

The cost of nutrient removal is the main concern in many places. Perhaps the principal caveat in regards to comparison of the costs of competing processes is to take care that the costs are developed on an equivalent basis. Costs must be compared as life cycle (i.e., present worth or equivalent annual) costs. Otherwise, the process that shifts the costs from a one-time capital expense to annual operating costs will appear to have an advantage when, in reality, it may not. In some cases, the goal may to be to minimize the construction costs, in which case the last statement may apply to a lesser degree.

7.1 Selection of Design Criteria

Design criteria can be obtained from any number of sources. The best and most reliable are based on the actual operation of similar full-scale facilities. Next desirable is the design criteria developed from a well-designed and

-operated pilot facility or a series of bench-scale tests in instances where there is an existing facility from which both comparable wastewater and mixed liquor are available. For processes where a significant amount of research and development and operational experience are available, it is common practice to select the design criteria from the technical literature. For certain types of processes or equipment, the best knowledge resides with the companies who market the process; however, great care must be taken before selecting a process on the basis of design criteria provided by a company (with or without a vested interest in the outcome of the evaluation). There are often irresistible pressures for a company to be overly optimistic in its choice of design criteria for a process that it markets or against which it competes. References of the current users of the technology should be checked thoroughly with phone calls and site visits, if possible, before accepting claims of the company.

7.2 Preliminary Sizing

The ideal situation would be to compare bids received on final design drawings and specifications for each process under detailed evaluation. This is seldom necessary or even desirable. Typically, it is adequate to prepare comparative cost estimates based on a preliminary design for each option. Cost estimates prepared from preliminary designs are greatly influenced by the assumptions made and the selection of the design criteria used to size the facilities. Again, care must be taken to be certain cost estimates have not been unrealistically biased in favor of a particular process as a result of the design criteria and cost data selected. Often, simple assumptions can have tremendous cost implications. Care should be exercised so that any assumptions made are reasonable and realistic. Reasonable care includes making preliminary investigations into site conditions, discussing the proposed facilities with local fire and building officials and other significant regulatory agencies, and adequately investigating any significant and unusual claims of equipment manufacturers.

7.3 Ability to Use Existing Assets

The use of existing assets is often a key consideration in many evaluations and, therefore, worth some additional discussion. In concert with the preliminary sizing, evaluation of nutrient removal technologies should consider the possibility of reusing existing assets that have remaining life. Often, the use of these assets will introduce inefficiencies in the system because they are not optimally sized for each process based on the preliminary sizing. However, the potential cost savings of using existing infrastructure may offset any inefficiency. The use of existing assets will often require additional effort in the planning and design phases as a relative percentage of construction costs, which should be considered as part of the evaluation.

7.4 Cost of Conservatism

When preparing preliminary designs for competing processes, it is critical, for an apples-to-apples comparison, to incorporate an equivalent safety factor or degree of conservatism to each design. For example, you would not compare one suspended growth nitrification process with a safety factor of 2 for nitrification with another that was costed with a nitrification safety factor of 1.

Valid reasons for using safety factors include the following:

- Account for unknowns in bacterial growth rates;
- Correct for reductions in bacterial growth rates caused by substrate limitations;
- Provide high efficiency of treatment;
- Ensure process stability; and
- Minimize risk of process failure because of low temperature, toxics, or influent mass loading variations.

It is important to be aware of the ways additional, and unnecessary, safety factors can be introduced to a design. This can occur by choosing overly conservative design temperatures or temperature correction coefficients and overly conservative kinetic coefficients, and using design wastewater strengths that have little similarity to the actual range of influent concentrations.

Overly conservative, and inconsistently applied, conservatism can cost a facility owner dearly at this stage of the design process by directing the selection to a process that is either not the most economical choice, or worse, one that will not perform properly under the actual conditions that the treatment facility experiences.

7.5 Historical Costs for Nutrient Reduction

Figures 5.12 and 5.13 present reported unit costs to remove nitrogen and phosphorus using various point-source control measures and BMPs for both agricultural and urban NPSs. Costs on a per-unit mass basis ($/kg or $/lb) to remove nutrients are site-specific and highly variable. The values reported in these two figures are for general information only and, as such, are not a substitute for doing the hard work of creating opinions of probable cost for specific projects. Numerous publications are available on methodologies

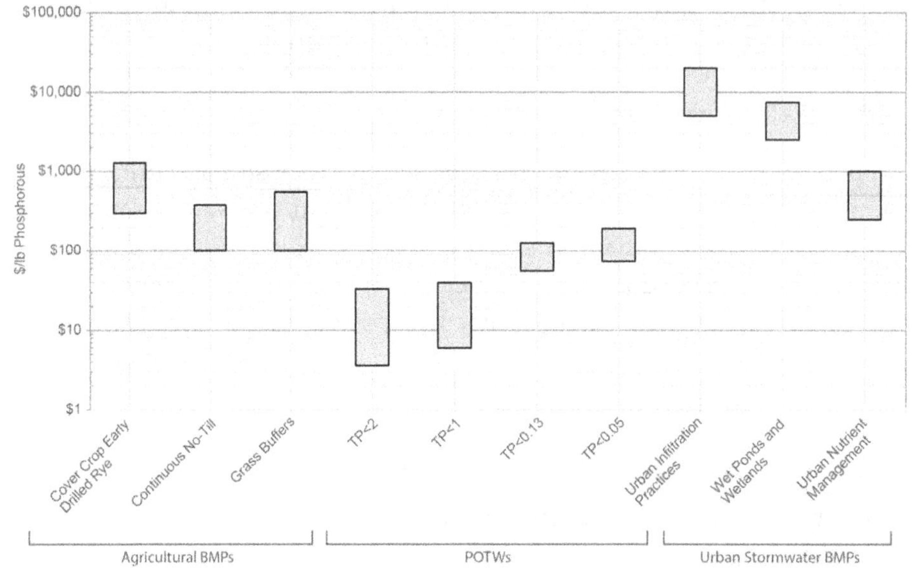

1. The sole purpose of this figure is to illustrate the reported ranges of unit costs for the removal of nitrogen or phosphorus. Many variables affect the unit costs for nutrient removal especially for BMPs and existing WRRFs.
2. Do not apply these unit values to specific facilities or watersheds. Where costs for the implementation of nutrient removal are required for planning and budgetary purposes, it is recommended that custom cost estimates be prepared based on appropriate level of detail.

FIGURE 5.12 Reported unit costs for phosphorus reduction (BMPs = best management practices; POTWs = publicly owned treatment works).

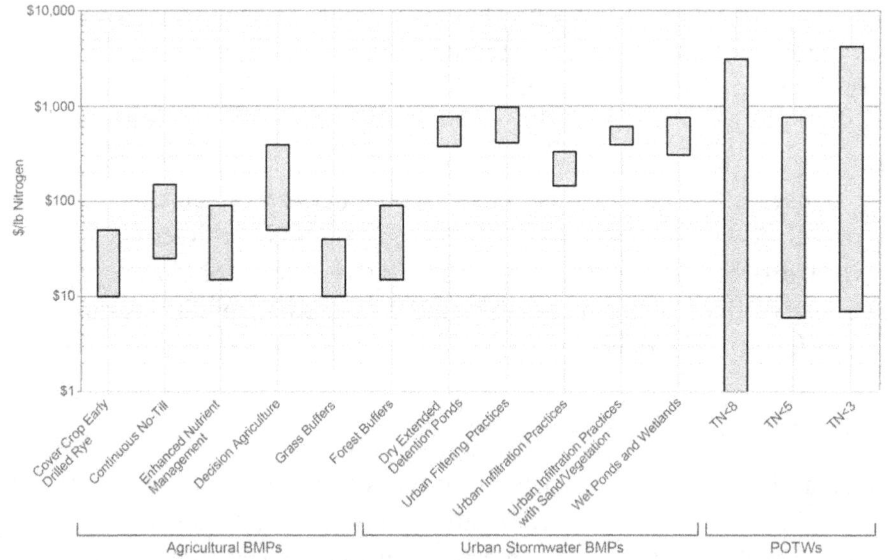

1. The sole purpose of this figure is to illustrate the reported ranges of unit costs for the removal of nitrogen or phosphorus. Many variables affect the unit costs for nutrient removal especially for BMPs and existing WRRFs.
2. Do not apply these unit values to specific facilities or watersheds. Where costs for the implementation of nutrient removal are required for planning and budgetary purposes, it is recommended that custom cost estimates be prepared based on appropriate level of detail.

FIGURE 5.13 Reported unit costs for nitrogen reduction (BMPs = best management practices; POTWs = publicly owned treatment works).

and requirements for cost estimating at various stages of a project (e.g., see those published by AACE International [http://www.aacei.org/], including AACEI [2011a, 2011b]).

The few studies that have evaluated costs to implement nutrient removal over a range of facility sizes for several levels of nutrient reduction have generally found that unit costs have increased with increasing restrictions on nutrient discharges (i.e., reduced design effluent nutrient concentrations) and decreased with increasing design treatment capacity (CH2M HILL, 2010; Jiang et al., 2005; Tetra Tech, 2011; Trowbridge, 2010).

Many factors induce the variability in these unit costs, including the availability of existing facilities and the feasibility of converting them to other uses; site constraints; geotechnical conditions; economic conditions; facility capacity; owner policies; measures included to improve site aesthetics (reduction of odor, noise, and visual effects); water quality limits (magnitude, frequency, and duration); degree of redundancy provided; and many others (WEF, 1998).

8.0 INNOVATION

Innovation is defined as a new idea, device, or method, or the act of introducing something new. Need typically drives innovation in the wastewater industry, primarily the desire to lower cost or improve performance. However, other benefits such as improved aesthetics or an increased ability to recover resources can provide the inspiration for innovation.

The well-known theory on the process by which innovations are introduced to practice (Rogers, 2003) applies equally well to the wastewater industry. Because of the long-term nature of many utility investments (often 20 years or more), the monopoly nature of utilities, the magnitude of investments in treatment facilities, the highly regulated operating environment, and the methods of funding, adoption of new technologies can take a long time, even when new technologies appear to offer significant savings or performance benefits. The tendency of human nature to resist change, water quality regulations, and the litigious nature of our society likely contribute to the hesitancy of many facility owners and design engineers to consider serious innovations in nutrient removal technologies. Water quality regulations hamper innovation because of their emphasis on technology-based standards, while society contributes to the hesitancy to innovate by its insistence on zero risk. In spite of the aforementioned conditions, innovation does have a place in the wastewater industry, and current economic conditions should see increasing incentives for facility owners to consider new concepts in nutrient removal.

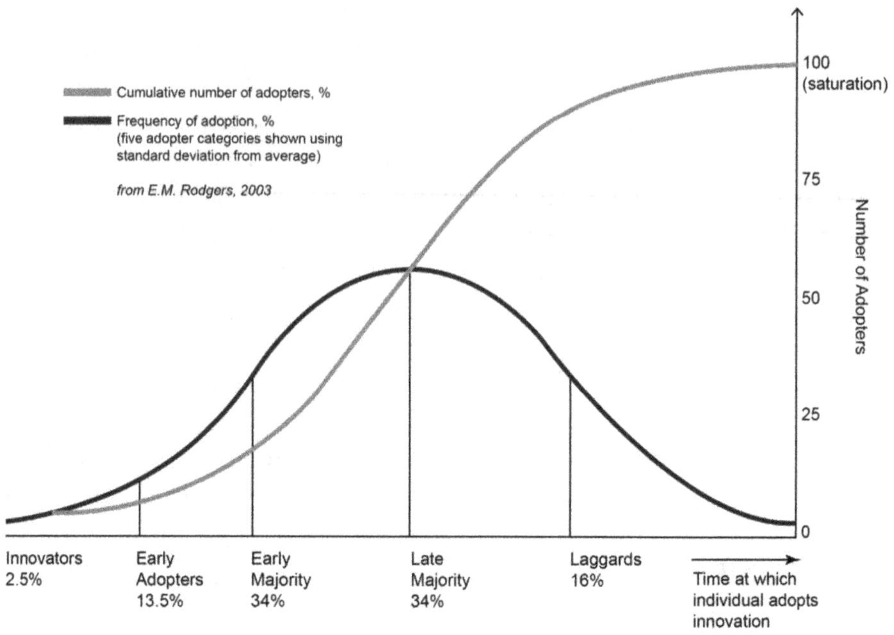

FIGURE 5.14 Evolutionary trends in technologies.

The wastewater industry has a long history of innovations that have served it well. This is illustrated by the technology timeline in Figure 5.15, which highlights some of the technological advances that have occurred since around 1850. The global challenges facing society, including increasing population and wealth, increasing demands on water resources, and the need to develop a sustainable society, ensure that advancements in nutrient removal technologies will continue at an increasing rate (van Loosdrecht, 2012). The challenge to utility owners and managers is to know when to implement innovative technologies.

The decision of whether to implement an innovative technology is a subjective one to be made by utility owners, with input from their advisors and regulators. Innovative technologies can present a number of risks, however, by making the effort to identify, investigate, and quantify the risks associated with a particular site and technology; owners can develop the knowledge needed to make an informed decision as to the probability of a successful implementation. Possible risks associated with innovative technologies include (1) process failure, primarily the inability to meet performance expectations; (2) budget overruns; (3) rejection by operating staff because of unforeseen labor requirements or undesirable working conditions; (4) equipment failure resulting from inadequate development or designs inconsistent with installation and operating conditions; and (5) lack

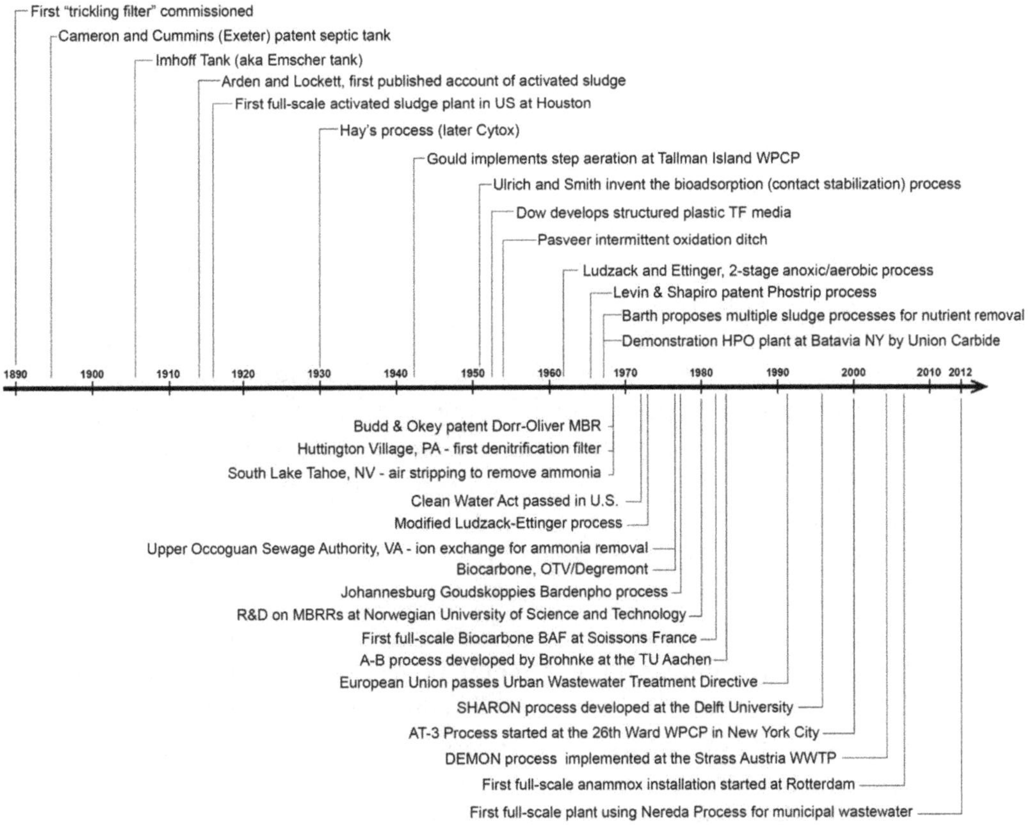

FIGURE 5.15 Timeline of selected innovations in wastewater treatment technology.

of needed redundancy or reliability resulting from a desire to make the new technology appear cost-competitive with established ones. Technologies that are progressively developed from inception through several generations of full-scale use will more often be successful than those that are rushed to market (Parker, 2011).

9.0 OTHER CONSIDERATIONS

9.1 Testing

Testing of BNR treatment technologies before selection or design is typically not required because of the large number of installed and successfully operating full-scale facilities treating municipal wastewaters. When wastewater characteristics are outside the norm, the technology is emerging or innovative or there is limited experience in removing a target constituent (e.g., refractory dissolved organic nitrogen); therefore, some type of testing of the treatment process is warranted. Treatability studies are typically categorized based on purpose and size as bench-scale, pilot-scale, or full-scale. Each type of test provides different types of information relative to the performance or cost of a facility. The cost to do any type of testing can be a good investment because it often reduces the cost of full-scale implementation by reducing uncertainty and associated risks, improving process performance, or reducing the size of the full-scale facility.

Bench-scale studies, including jar and bucket tests, provide information on process kinetics and water chemistry and can help set the direction for further larger-scale tests, as required. Bench tests have the advantage of being relatively quick and inexpensive to accomplish and providing information for calibration of process models; however, bench tests require representative samples of the wastewater and an acclimated biomass. In addition, the results can be difficult to scale and cannot provide information on costs, mechanical equipment, or operating requirements.

Pilot testing requires a larger investment of time and money and requires larger volumes of the wastewater to be tested, but can yield information on the response of the process to variations in flow and quality of wastewater, design criteria, operations and maintenance requirements, and more definition on the expected cost for a full-scale facility. In addition, pilot testing allows direct comparisons of competing technologies. The bulk of the cost associated with pilot testing is the labor and analytical work, although the cost of pilot-plant facilities can be significant.

Full-scale demonstration testing of potential process modifications may be possible at existing facilities, and the associated costs can be less than those for pilot testing. Compared to bench and pilot testing, full-scale demonstrations eliminate the uncertainty associated with scale-up; however, the flexibility seldom exists to test multiple process and operating conditions. With demonstration-scale testing, there is also concern over the ability to meet permit limits during test periods. A number of text and reference books contain additional information on treatability studies (Cleary, 2014; Grady et al., 2011; McEwen, 1998).

9.2 Mathematical Models

Mathematical process models theoretically provide a method to evaluate the expected performance of many alternative processes quickly and easily. Current biological treatment models, however, require the input of many kinetic coefficients and influent wastewater characteristics that are not commonly or easily measured. Although the models are not sensitive to the value of many of the input coefficients, there is enough sensitivity to a limited number of the parameters to create some uncertainty about the accuracy of model predictions without calibration and verification of key coefficients.

9.3 Patents

Nutrient removal processes, especially new and innovative processes, are often covered by U.S. patents. Patent holders typically require payment of a fee for the use of the patented technology or the sole-source purchase of the equipment used in the patented process. During the selection of a nutrient removal process, the costs of a process patent are simply another factor to consider. If a treatment process is the most economical alternative, or otherwise best meets the selection criteria, the presence of a process patent in itself should not be cause to eliminate the process from consideration.

Unfortunately, process patents many times become marketing tools used by equipment manufacturers and their representatives. Common ploys are to not charge a fee if certain items of equipment are sole sourced, or to use the patent to create a large "package" of equipment that may be offered to contractors as an "all or nothing" deal. Other problems can arise when two suppliers claim to have valid patents for the same process, or when a patent has expired and the patent holder still asks for a license fee. In such cases, the advice of a patent attorney can be invaluable. In any event, the aforementioned problems can typically be dealt with simply and economically, provided the owner is cognizant of the potential pitfalls in advance.

10.0 THE LAST STEP—THE FINAL CHOICE

The final choice of which treatment technology to implement is ultimately a policy decision to be made by the owner after receiving input from engineers, staff, and possibly ratepayers and the environmental community. It is often useful to prepare simple summary tables of the principal advantages and disadvantages of the short-listed processes. These serve to help the decision-makers focus on the results of the evaluation process, relate any numerical ranking to concrete values, and put the cost comparisons in perspective. For significant projects, value engineering also can provide an independent opinion and some extra comfort that the right choice is being made.

It cannot be emphasized enough that the final choice is a subjective value judgment that must reflect the things of most importance to the people that will own, operate, and maintain the treatment facility and will be affected by its effect on the water quality of the receiving waterbody.

11.0 FOLLOW THROUGH

Nutrient removal systems are more complex than conventional treatment systems and are typically less forgiving of mechanical failures or operator inattention because they must meet tighter effluent limits. Therefore, during technology selection, it is important to plan for the staffing required to operate and maintain the systems. It is also important to establish a detailed plan for facility operation including startup and optimization.

11.1 Training

Staff training is critical for successful implementation of a nutrient removal program. Facility operations and maintenance staff must be trained in all aspects of a new treatment system. Involving operations and maintenance staff throughout the entire life of the project is important.

A successful training program engages staff during planning and through the design, construction, and commissioning of the new treatment facilities. Training programs are site specific and may vary from workshops and manufacturer presentations to full certification programs. Minimally, training programs may include the following:

- Basic mechanisms of nutrient removal,
- Monitoring parameters,
- Key operating parameters,
- Key performance indicators,
- Equipment,
- System controls,
- Startup training, and
- Troubleshooting.

11.2 Startup and Testing

Operations and maintenance staff has the first hands-on contact with a new treatment system during startup and testing. Therefore, startup is important not only for testing and verification of the new systems, but also for training of operation and maintenance staff. Startup procedures must be carefully established during planning and design phases of the project. All startup activities must be thoroughly described in the project specifications. The following is a suggested startup approach:

- The design engineer prepares a general startup plan describing objectives and main startup activities;
- This general plan is then used by the contractor to produce a detailed startup plan, including testing of each piece of equipment, testing of the new system as a whole for a preestablished period of time, and startup schedule;

- Startup activities are ideally conducted with the participation of design engineer, owner's representatives (operations and maintenance staff), and the contractor; and
- Verification of metrics and approval are conducted by inspection personnel and the owner's representatives.

11.3 Optimization

Process optimization can be done after or even during startup to improve performance of a new treatment system. Optimization during startup can be more challenging because multiple variables may be changing at the same time, and equipment failure is more common. Ideally, optimization should take place after startup and throughout the life of the facility. Optimization can target a portion of the treatment system or the system as a whole. For example, optimization of a mixed liquor recycling system by addition of flow metering capabilities can be considered as optimization of a portion of the system. On the other hand, optimization of the treatment system by addition of sidestream treatment involves the system as a whole. Strategies for optimization are case specific. The following traditional "plan–do–check–adjust" approach can be used as a general basis for an optimization program, as follows:

- Plan—define current condition, establish goals, identify roadblocks, identify key process operation parameters, identify optimization opportunities, run process model, and develop optimization actions;

1. Do—implement optimizations actions;
2. Check—measure performance and identify issues;
3. Adjust—adjust key process operating parameters and standardize successful optimization measures; and
4. Back to plan—run the cycle again.

12.0 REFERENCES

AACE International (2011a) *AACE International Recommended Practice No. 17R-97—Cost Estimate Classification System*. Rev. November 29, 2011 ed.; AACE International: Morgantown, West Virginia.

AACE International (2011b) *AACE International Recommended Practice No. 18R-97—Cost Estimate Classification System for the Process Industries*. Rev. November 29, 2011 ed.; AACE International: Morgantown, West Virginia.

American Society of Civil Engineers; U.S. Environmental Protection Agency; Water Environment Research Foundation; Federal Highway Administration; American Public Works Association (2015) International Stormwater Best Management Practices (BMP) Database. http://www.bmpdatabase.org/ (accessed Feb 2015).

Borah, K.; Bera, M. (2003) Watershed Scale Hydrologic and Nonpoint Source Pollution Models: Review of Mathematical Bases. *Trans. ASAE*, 46, 1553–1566.

Chesapeake Bay Commission (2004) *Cost-Effective Strategies for the Bay—Six Smart Investments for Nutrient and Sediment Reduction*; Chesapeake Bay Commission: Annapolis, Maryland.

Coats, E. R.; Dobroth, Z. D.; Brinkman, C. K. (2015) EBPR Using Crude Glycerol: Assessing Process Resiliency and Exploring Metabolic Anomalies. *Water Environ. Res.*, 87, 68–79.

Droste, R. L. (2009) *Theory and Practiceo of Water and Wastewater Treatment*; Wiley & Sons: New York.

Grady Jr., C. P. L.; Daigger, G. T.; Love, N. G.; Filipe, C. D. M. (2011) *Biological Wastewater Treatment*, 3rd ed.; IWA Publishing/CRC Press: Boca Raton, Florida.

Henze, M.; Harremoes, P.; la Cour Jansen, J.; Arvin, E. (2002) *Wastewater Treatment: Biological and Chemical Processes*, 3rd ed.; Springer-Verlag: Berlin, Germany.

Henze, M.; van Loosdrecht, M. C. M.; Ekama, G. A.; Brdjanovic, D. (2008) *Biological Wastewater Treatment: Principles, Modeling, and Design*; IWA Publishing: London, U.K.; p 528.

Horgan, C. J.; Coats, E. R.; Loge, F. J. (2010) Assessing the Effects of Solids Residence Time and Volatile Fatty Acid Augmentation on Biological Phosphorus Removal Using Real Wastewater. *Water Environ. Res.*, 82 (3), 216–226.

Jimenez, J.; Dursun, D.; Dold, P.; Bratby, J.; Keller, J.; Parker, D. (2010) Simultaneous Nitrification-Denitrification to Meet Low Effluent Nitrogen Limits: Modeling, Performance and Reliability. *Proceedings of the 83rd Annual Water Environment Federation Technical Exhibiton and Conference* [CD-ROM]; New Orleans, Louisiana, Oct 2–6; Water Environment Federation: Alexandria, Virginia.

McAllister, D. M. (1982) *Evaluation in Environmental Planning—Assessing Environmental, Social, Economic, and Political Trade-Offs*; The MIT Press: Cambridge, Massachusetts.

Merriman, K. R.; Gitau, M. W.; Chaubey, I. (2009) A Tool for Estimating Best Management Practice Effectiveness in Arkansas. *Appl. Eng. Agric.*, **25** (2), 199–213.

Munn, M. D.; Moran, P.; Alexander, R. B.; Dubrovsky, N. M.; Hoos, A. B.; Woodside, M. D.; Short, T. M. (2009) The U.S. Geological Survey National Water-Quality Assessment Program–Advances in Understanding Sources, Transport, and the Biological Effects of Nutrients from Headwater Streams to Coastal Rivers (Poster): Fifth Symposium on Harmful Algae in the U.S.; Nov 16, 2009.

North Carolina Department of Environment and Natural Resources (2014) *North Carolina Nonpoint Source Pollution Management Program, Third Update*; North Carolina Department of Environment and Natural Resources, Division of Water Resources, Water Planning Section: Raleigh, North Carolina.

Oehmen, A.; Lemos, P. C.; Carvalho, G.; Yuan, Z. G.; Keller, J.; Blackall, L. L.; Reis, M. A. M. (2007) Advances in Enhanced Biological Phosphorus Removal: From Micro to Macro Scale. *Water Res.*, **41**, 2271–2300.

Parker, D. S. (2011) Introduction of New Process Technology into the Wastewater Treatment Sector. *Water Environ. Res.*, **83**, 483–487.

Rogers, E. M. (2003) *Diffusion of Innovations,* 5th ed.; Simon and Schuster: New York.

Srivastava, M. M.; Sanghi, R. (2005) *Chemistry for Green Environment*; Alpha Science International: Oxford, U.K.

Tchobanoglous, G.; Stensel, H. D.; Tsuchihashi, R.; Burton, F. (2014) *Wastewater Engineering: Treatment and Resource Recovery,* 5th ed.; Metcalf & Eddy/AECOM: New York.

U.S. Environmental Protection Agency (2003) *National Management Measures to Control Nonpoint Source Pollution from Agriculture*; EPA-841/B-03-004; U.S. Environmental Protection Agency: Washington, D.C.

U.S. Environmental Protection Agency (2005a) *Handbook for Managing Onsite and Clustered (Decentralized) Wastewater Treatment Systems*; EPA-832/B-05-001; U.S. Environmental Protection Agency: Washington, D.C.

U.S. Environmental Protection Agency (2005b) *National Management Measures to Control Nonpoint Source Pollution from Forestry*; EPA-841/B-05-001; U.S. Environmental Protection Agency: Washington, D.C.

U.S. Environmental Protection Agency (2005c) *National Management Measures to Control Nonpoint Source Pollution from Urban Areas*;

EPA-841/B-05-004; U.S. Environmental Protection Agency: Washington, D.C.

U.S. Environmental Protection Agency (2005d) *National Management Measures to Protect and Restore Wetlands and Riparian Areas for the Abatement of Nonpoint Source Pollution*; EPA-841/B-05-003; U.S. Environmental Protection Agency: Washington, D.C.

U.S. Environmental Protection Agency (2007a) *Biological Nutrient Removal Processes and Costs*; EPA 823-R-07-002; U.S. Environmental Protection Agency: Washington, D.C.

U.S. Environmental Protection Agency (2007b) *National Management Measures to Control Nonpoint Source Pollution from Hydromodification;* EPA-841/B-07-002; U.S. Environmental Protection Agency: Washington, D.C.

U.S. Environmental Protection Agency (2009) *The Next Generation of Tools and Actions to Restore Water Quality in the Chesapeake Bay*. U.S. Environmental Protection Agency: Washington, D.C.

U.S. Environmental Protection Agency (2010a) *Guidance for Federal Land Management in the Chesapeake Bay Watershed*; EPA-841-R-10-002; U.S. Environmental Protection Agency: Washington, D.C.

U.S. Environmental Protection Agency (2010b) *Nutrient Control Design Manual*; U.S. Environmental Protection Agency: Washington, D.C.

U.S. Environmental Protection Agency (2013a) *Emerging Technologies for Wastewater Treatment and In-Plant Wet Weather Management*; EPA-832/R-12-011; U.S. Environmental Protection Agency: Washington, D.C.

U.S. Environmental Protection Agency (2013b) Watershed and Water Quality Modeling Technical Support Center. http://www.epa.gov/athens/wwqtsc/index.html (accessed Feb 2015).

U.S. Environmental Protection Agency (2014) Polluted Runoff: Nonpoint Source Pollution. http://water.epa.gov/polwaste/nps/ (accessed Feb 2015).

U.S. Environmental Protection Agency (2015) EPA Center for Exposure Assessment Modeling (CEAM). http://www2.epa.gov/exposure-assessment-models (accessed Feb 2015).

van Haandel, A. C.; van der Lubbe, J. G. M. (2012) *Handbook of Biological Wastewater Treatment Design and Optimisation of Activated Sludge Systems,* 2nd ed.; IWA Publishing: London, U.K.

Versprille, A. I.; Zuurveen, B.; Stein, T. (1984) The A-B Process: A Novel Two Stage Wastewater Treatment System. *Water Sci. Technol.*, **17**, 235–246.

Water Environment Federation (1998) *Biological and Chemical Systems for Nutrient Removal Special Publication*; Water Environment Federation: Alexandria, Virginia.

Water Environment Federation (2010) *Nutrient Removal;* WEF Manual of Practice No. 34; McGraw-Hill: New York.

Water Environment Federation; American Society of Civil Engineers; Environmental and Water Resources Institute (2005) *Biological Nutrient Removal (BNR) Operation in Wastewater Treatment Plants;* WEF Manual of Practice No. 30; McGraw-Hill: New York.

Water Environment Federation; American Society of Civil Engineers; Environmental and Water Resources Institute (2009) *Design of Municipal Wastewater Treatment Plants,* 5th ed.; WEF Manual of Practice No. 8, ASCE Manuals and Reports on Engineering Practice No. 76; McGraw-Hill: New York.

6

Identifying and Managing Risks

1.0 INTRODUCTION

Nutrients pose significantly different risks than pathogens, toxicants, and conventional pollutants, which are the primary aim of most water resource recovery facilities (WRRFs). Because nutrients are required for the growth and health of all organisms, they cannot be completely eliminated from the natural environment. Long before modern sanitation and treatment practices became widespread, domestic wastes were typically returned to the natural environment, sometimes intended for the beneficial recovery of nutrients (Ashley et al., 2011). Analogous practices continue to be used by some utilities today (e.g., biosolids land-application programs and irrigation with effluent).

In addition to minimizing waterborne human health risks and aquatic toxicity, utilities of the future should consider their ability to help sustainably manage nutrients for beneficial reuse and to control their release at rates and conditions that best match the needs of the natural environment. Some of the risks involved in these new endeavors will be unique to nutrients, while others will be similar to those more routinely encountered. The considerations discussed in the remainder of this chapter can be used by project and program managers to develop registers and matrices (Figure 6.1) to communicate and assess the risks and rewards of opportunities and alternatives.

2.0 SITE-SPECIFIC ECOSYSTEM RESPONSE

The response of a receiving waterbody to nutrient loadings is site-specific and is generally a complex phenomenon that is controlled by the physical, chemical, and biological properties of each individual system (see Chapter 3 for further details). Understanding the ecosystem load response for each particular scenario is critical. For this reason, an assessment of the effects

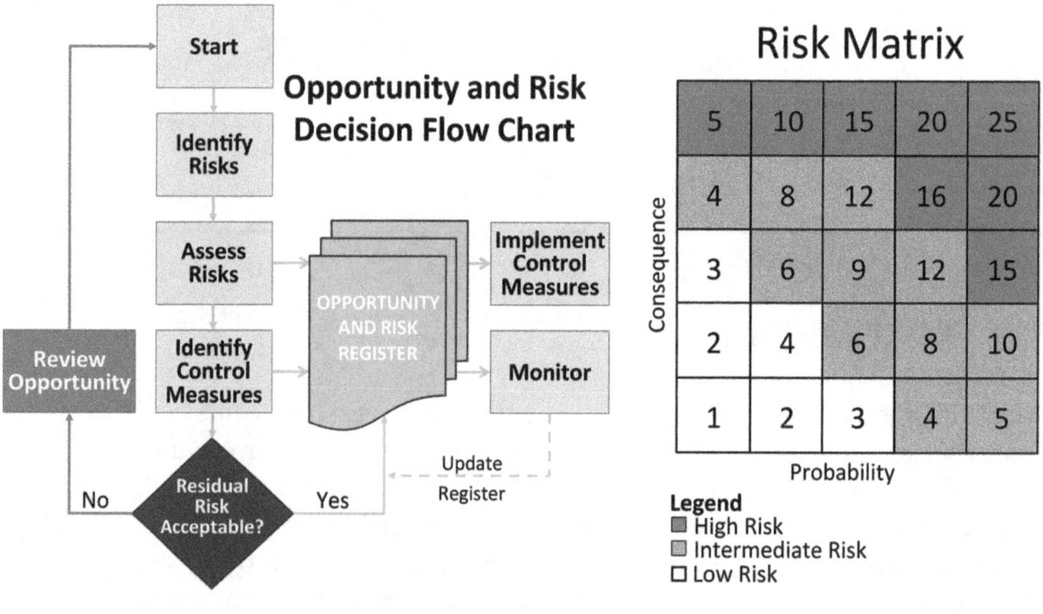

FIGURE 6.1 A risk register and matrix are just a few of the tools that can be used to help quantify and manage risks and also help communicate both opportunities and risks for different project and program elements.

of nutrient loading in each receiving waterbody, made through a continuous monitoring program, is necessary to determine the ecosystem response to nutrient inputs. Results from continuous monitoring may be incorporated to computer-based models to help managers develop effective policies and implement optimized solutions for each watershed. This understanding will further help establish nutrient load parameters that will protect and improve the water quality in streams, lakes, and reservoirs while potentially minimizing costs for nutrient removal and/or recovery.

Lakes and reservoirs around the world experience accelerated eutrophication. Although the causes of eutrophication are complex and vary for different aquatic systems, it is widely known that nutrient enrichment exacerbates this process. A wide range of strategies have been developed to decrease nutrient loads to surface waters, and the historical trend in many of these management strategies has been to focus on decreasing external nitrogen and phosphorus loads by placing increasingly stringent effluent limits on WRRF discharges or by controlling nonpoint sources. However, nutrient limits do not need to be cast in stone. Ecosystems are constantly changing, and not all of them will benefit from overly stringent nutrient limits all the time. Depending on the ecological structure of the watershed and management goals, having the ability to change WRRF nutrient limits to match the changing needs of the ecosystem may be part of best management practices for water quality protection.

2.1 Phosphorus Load Reduction Uncertainties

Phosphorus is a critical limiting nutrient in an aquatic ecosystem that exacerbates eutrophication if present in excessive amounts. When analyzing the fate of phosphorus in aquatic systems, it is important to consider the interdependent relationship between phosphorus and nitrogen. Phosphorus loading into an aquatic system may occur in two different forms. The first one is external loading from point and nonpoint sources. The second form is known as *internal phosphorus loading,* and may be a substantial fraction of the total (Lake et al., 2007). Internal loading includes the release of phosphorus from deposited sediments. The causes of sediment phosphorus release are numerous and complex, but have been principally attributed to the anaerobic microbial reduction of iron oxide complexes containing phosphates. Under reduced conditions (anaerobic), phosphate is released into the water column, contributing to the excessive production of autotrophs, especially algae and cyanobacteria. Nitrate in surface waters maintains a well-oxidized environment that promotes phosphorus adsorption (mainly orthophosphate) to ferric iron in the sediments, thereby preventing phosphorus from becoming readily available to aquatic biota. Therefore, nitrate loading to surface waters may extensively decrease internal phosphorus loading (Gabriel et al., 2006; Petzoldt and Uhlmann, 2006).

2.2 Nitrogen Load Reduction Uncertainties

Nitrogen stimulates phytoplankton productivity and can exacerbate eutrophication. In addition, some nitrogen species, under certain conditions, can be toxic to aquatic organisms and humans. The general consensus from a lake and reservoir management perspective is that reducing nitrogen loads to an aquatic system is always beneficial. However, this conclusion is not always valid and should be tempered with a deeper understanding of the unique properties of different nitrogen species such as nitrate. There are situations where nitrate discharged to streams, lakes, and reservoirs actually improves water quality (Cubas et al., 2014). In these instances, nitrate helps prevent undesirable anaerobic conditions in the benthos and increases alkalinity when denitrification occurs. Furthermore, high nitrate loads help maintain a high nitrogen-to-phosphorus ratio that discourages the proliferation of cyanobacteria, thus helping to control harmful algal blooms.

Nitrate discharged from a WRRF may serve as an efficient method to improve the water quality in receiving waterbodies. Benefits from high nitrate loads will vary depending on the physical, chemical, and biological properties of the particular system. The properties of each system will control nitrogen removal rates by providing an adequate environment for denitrification or for nitrogen burial or consumption. When suitable environmental conditions prevail for nitrate removal in surface waters, a WRRF should be able to temporarily shut down or modify its denitrification process and discharge a nitrified product water to the receiving waterbody. In some cases, this may help water utilities decrease operational costs associated with nitrogen removal; however, nitrification and denitrification processes should be evaluated on a case-by-case basis to optimize operations and determine if supplemental alkalinity will be required. While scientific debate continues over the potential advantages and disadvantages of nitrate discharges, site-specific water quality studies are recommended to assess the consequences of implementing it as a management strategy for a particular waterbody.

3.0 REGULATORY FRAMEWORK

Nutrient enrichment is recognized as a pressing problem worldwide. While various countries have different regulatory frameworks aimed at managing nutrients to produce acceptable water quality, this roadmap focuses on the regulatory framework in the United States. Table 6.1 summarizes the most

TABLE 6.1 Key federal water quality permitting statutes and regulations (CWA = Clean Water Act; TMDL = Total Maximum Daily Load; and NPDES = National Pollutant Discharge Elimination System).

Statute/ regulation	Legal citation	Function
Statute - CWA	33 U.S. Code Chapter 26	Establishes a framework with the intent of restoring and protecting the quality of the nation's waters
§301	33 U.S. Code § 1311	Authority to establish permit limitations
§302	33 U.S. Code § 1312	Authority to establish water quality-based effluent limitations
§303(a) through (c)	33 U.S. Code § 1313(a) – (c)	State authority to develop, adopt, and revise water quality standards
§303(d)	33 U.S. Code § 1313(d)	Evaluation of impairment and establishment of TMDLs
§402	33 U.S. Code § 1342	Establishment of NPDES
Regulation	**40 CFR Chapter I Subchapter D**	**Implementation of the CWA**
§125	40 CFR §125	Implementation of criteria and standards for NPDES Permits
§130.7	40 CFR §130.7(d)	TMDL and individual water quality-based effluent limitations

pertinent portions of the Clean Water Act (CWA) statute and its implementing regulations. Many of these regulatory requirements were introduced in Chapters 2 and 3, while this chapter focuses on potential compliance risks and concepts for mitigating those risks.

Nutrient reduction from WRRFs in the United States is ultimately governed under the National Pollutant Discharge Elimination System (NPDES) regulatory scheme promulgated in the CWA. The NPDES scheme has served the United States well through its first three decades, and led to cleaner water at a cost that has generally been affordable to the public. Point-source control of pollutants almost exclusively solved most of the nation's water quality problems with little reliance on nonpoint source control. Effective nutrient management, however, confounds the traditional NPDES scheme. Clear relationships between surface water impairment and nutrient concentrations are often difficult to identify, have numerous underlying variables, and typically differ from waterbody to waterbody. In addition, unregulated nonpoint sources are often the largest overall contributors of nutrients and must be included in any successful nutrient reduction plan. Given its inherent complexities and uncertainties, a new regulatory model is needed to successfully address nutrients. Specifically, a model is needed that embraces an adaptively managed watershed-based approach allowing WRRFs and regulators to minimize both environmental and financial risk.

Adaptive management is a process that acknowledges uncertainty by taking initial actions to address a problem, learning from the initial outcome, then adjusting actions, if needed, to reach a desired outcome. Adaptive approaches tend to be cyclical or iterative and span longer time periods compared to linear approaches. They have been successfully used by regulatory agencies to manage natural resources (wildlife, fisheries, forests, etc.). Therefore, applying adaptive management to water resources is a logical progression. The precautionary principle is used extensively in Europe to address uncertainty and seeks to ensure actions taken to address environmental problems are cost-effective based on the presumed risk. Combining these two concepts allows nutrient pollution to be addressed in a sustainable manner that minimizes environmental and financial risk by taking proactive, proportionate steps.

Regulators need to be careful to balance the possible unintended consequences from moving forward too fast with overly stringent point-source limits (e.g., anti-backsliding rules) and financial consequences associated with removing nutrients from point and nonpoint sources with the possibility of further affecting ecological integrity by doing nothing. Because nitrogen removal can be much more costly and energy-intensive than phosphorus removal, "one-size-fits-all" technology-based effluent limitations (TBEL) for total nitrogen may not always be appropriate. Some cases are best suited to a site-specific approach such as water quality-based effluent limitations (WQBEL), for instance, when a waterbody may be affected by ammonia toxicity and/or a nitrate drinking water criterion or when high nitrogen loads negatively affect native aquatic biota such as sensitive macrophytes or mollusks (Benndorf and Chorus, 2006). In other cases, TBEL may be appropriate and beneficial (e.g., such as instances where nonpoint sources of nitrogen overwhelm a waterbody).

As mentioned in Section 2.2, work by Cubas et al. (2014) found that controlled nitrate discharges from a WRRF to the Occoquan Reservoir actually lowered the potential for harmful algal blooms and contributed to improved reservoir water quality and lower nutrient loads in releases from the reservoir. Thus, for freshwater impoundments, WRRF managers should seriously question stringent total nitrogen limits that may unintentionally do more harm than good to the aquatic environment, especially if the limits are so low that additional energy or carbon supplementation would be required that may negatively affect the carbon footprint. For flowing fresh water, downstream effects of nitrogen loads, such as impoundments, estuaries, or marine waters, need to be taken into account.

Because of the regulatory nature of nutrient management strategies, those strategies will have inherent risk for point sources. Minimization of risk will lead to more successful implementation of nutrient controls. Risks include noncompliance with regulatory objectives set for receiving bodies of water including streams, lakes, reservoirs, estuarine systems, and marine waters. This is particularly true when permit limits are involved. As discussed previously, it is difficult to establish numeric water quality standards for nutrients. Once established, however, they can result in low permit limits for WRRFs, even those that may be a small contributor to the overall waterbody nutrient load. This is where a total maximum daily load study (TMDL) may be the best tool to adaptively manage nutrient reduction in a watershed (see Chapter 2 for further details on TMDLs). A TMDL can establish a wasteload allocation that relies on technology-based limits (or some other limit that is less stringent than water quality-based permit limits), while establishing equitable reduction targets for nonpoint sources. The TMDL provides a regulatory mechanism that allows for time to evaluate water quality improvements. If water quality goals are achieved, no further reductions are necessary. If water quality goals are not achieved in a period of time, the TMDL can be adaptively managed to look at additional incremental controls over an additional period of time. An example of a phased, adaptive management TMDL approach for phosphorus is the Big Creek, Kansas, TMDL.

Regardless of whether NPDES permit limits are established by a TMDL or through WQBELs, they must be reasonable and attainable within the fiscal constraints of the permittee. In addition to numeric limits expressed in terms of magnitude, the duration of the limit is also crucial. Nutrients rarely cause short-term effects; as such, permit limits should be expressed as long-term averages to mimic the effect of eutrophication, which is typically a cumulative effect over a long period of time (seasonal or annual). Because eutrophication is the dominant environmental driver, limits based on long-term averages are most practical for WRRFs. Long-term averages are even more important to biological nutrient removal (BNR) processes that typically perform best under relatively steady-state conditions. Limits expressed as overly stringent short-term averages (daily, weekly, or even monthly) may actually discourage sustainable practices or require short-term changes that may have no true environmental benefit. If short-term limitations are deemed necessary for protection of acute water quality criteria, WQBELs should be developed based on acute criteria (as opposed to chronic criteria) and should specify not only magnitude, but also duration and frequency of allowable excursions. Work by Lindeke (2014) describes the use of a rolling annual average to control total phosphorus discharges from an enhanced biological phosphorus removal (EBPR) facility and potential compliance risks if monthly averages were used instead.

In summary, the inherent complexities and uncertainties of nutrient management confound traditional pollutant control methods. Therefore, regulatory control of nutrient point sources needs to move deliberately and adaptively, while allowing for limitation changes as more is learned about nutrients and their effect on the environment and focusing regulatory attention where it will provide the most ecological benefit.

4.0 FINANCIAL RISKS

As with other WRRF programs and projects, nutrient management requires sustainable funding mechanisms for day-to-day operations and maintenance and capital for periodic investments in treatment infrastructure upgrades, rehabilitation, and equipment replacement (see Chapter 2 for details on various funding vehicles). The financial risks for nutrient management are somewhat similar to those encountered for conventional pollutants; however, different facilities, technologies, and operations will likely be required and are discussed further elsewhere in this publication. Potential rewards for these financial risks should also be considered, along with the goal that nutrient management be cost neutral for the watershed. Considerations include the following:

- Environmental credits—monetizing environmental effects (both positive and negative) of nutrient removal and recovery alternatives can help quantify aspects traditionally evaluated through noneconomic methods, such as greenhouse gas emissions, carbon footprint, and effects to waterbody and riparian flora and fauna. This can help demonstrate the value of different nutrient management scenarios to community leaders and ratepayers;
- Watershed nutrient trading—nonpoint sources and other WRRFs in the same watershed should be evaluated to determine the feasibility of pooling funds to help finance the most economical solution for the watershed. For instance, similar or better environmental benefit may be gained by contributing funds toward BNR upgrades at a large WRRF or widespread nonpoint-source control than to finance similar upgrades at smaller WRRFs;
- Resource recovery—besides the removal of nutrients from WRRF liquid streams, some alternatives involve nutrient recovery. Furthermore,

some nutrient recovery alternatives require anaerobic digestion, opening up the potential for biogas-to-energy opportunities. Marketplace supply versus demand is a key issue that may affect the business case for adopting resource recovery. Markets for the recovered resources along with market competition from other products should be evaluated, in addition to the potential risks if the cost of recovery exceeds revenue; and

- Stringent limits—beyond a particular threshold, effluent nutrient limits tend to pose diminishing returns on investment. In some cases, utilities should consider reuse alternatives, particularly in situations where effluent can be beneficially used for nearby agriculture, tree nurseries, and landscaping irrigation. In some areas, whole effluent or liquid biosolids reuse may be feasible alternatives (pending local land use and other regulatory restrictions).

WRRFs should consider the higher capital and operating costs for nitrification compared to carbonaceous-only removal. Some of the higher operating costs can be recouped by good design and operation of denitrification processes. For nitrifying WRRFs with anaerobic digestion, sidestream deammonification offers potentially significant operational savings associated with lower oxygen, carbon, and alkalinity demand. While deammonification is rapidly becoming an established alternative for digestate dewatering liquors, the slow growth rate of anaerobic ammonium oxidation bacteria, competition with nitrite-oxidizing bacteria, and the need for more advanced instrumentation and process control are among the challenges to its feasibility and widespread application to the main liquid treatment train. Significant research in this area includes an ongoing project by the Water Environment Research Foundation (2014). If mainstream deammonification can be made to work reliably, it could revolutionize the WRRF industry in terms of nitrogen control and energy neutrality.

Compared to nitrogen removal, phosphorus removal is generally less costly. Enhanced biological phosphorus removal may require more capital investment compared to chemical alternatives, but it can offer significant operational savings and open up avenues for phosphorus recovery that should be considered. Chemical precipitation alternatives typically require lower capital investment; however, the ongoing operations and maintenance costs of chemical addition and the effects to downstream liquid and solids treatment processes must be considered. Selection of the optimal alternative for a given WRRF requires site-specific comprehensive life cycle assessments.

For WRRFs that practice EBPR and anaerobic digestion, predigestion phosphate stripping from waste activated sludge (WAS) offers potentially significant operational and maintenance savings by preventing nuisance struvite deposits and improving biosolids dewaterability, although this latter benefit represents an area of ongoing research (Shimp et al., 2014). When paired with a struvite recovery system, additional operating cost benefits should be considered because of lowered phosphate and ammonia recycle loads to the liquid stream, reduced phosphorus content in biosolids, and potential revenue from the struvite fertilizer byproduct. Revenue generation from a commercial byproduct at a WRRF may be a new concept for some utilities and will likely require further investigation and new business practices and procedures. Struvite recovery and sidestream deammonification should not be considered mutually exclusive alternatives; in fact, sidestream struvite recovery in combination with deammonification should continue to be researched, developed, and evaluated for WRRFs that practice anaerobic digestion.

Utilities should also consider economies of scale. Some technologies may require a centralized regional facility to be commercially feasible. For example, a regional biosolids facility may be the most economical solution for biofuel and nutrient recovery in areas that have multiple small- to medium-scale liquid stream treatment facilities.

5.0 TREATMENT PROCESS RISKS

Nutrient management at a WRRF generally involves removing nutrients from the liquid stream and then recovering them from the more nutrient-rich biosolids stream. In many instances, the most sustainable alternatives involve BNR; however, there are misconceptions regarding BNR that can steer planning and design decisions in the wrong direction. For example, EBPR and denitrification have been thought by some to be somewhat mutually exclusive and unreliable for cold or temperate climates. However, work by Baeza et al. (2014) helps explain how the two processes can coexist, and many facilities have overcome these misconceptions by including carbon supplementation or fermentation to provide consistent and reliable performance (Bunch et al., 2014; Fitzpatrick et al., 2014). That being said, as with any WRRF process, the potential for upsets or substandard performance should be considered in BNR design and operation.

5.1 Changes to Influent Characteristics

Influent characteristics are the biggest determining factor for nutrient removal performance. As more facilities move toward the recovery of nutrients and other resources from wastewater streams, the need to properly characterize wastes will become increasingly important. Knowing the average concentration of biochemical oxygen demand alone is not enough. Optimal design requires knowledge of the influent carbon (often expressed as chemical oxygen demand [COD]) and phosphorus and nitrogen fractions (particulate, colloidal, truly soluble, readily biodegradable, non-biodegradable, etc.) and their variability because of source changes, temperatures, and infiltration and inflow (I/I) (Melcer et al., 2003). High denitrification rates require sufficient soluble biodegradable carbon, and EBPR requires a mixture of volatile fatty acids (VFAs). Conventional physical, chemical, and BNR methods are not able to remove certain fractions of nitrogen and phosphorus, known as *recalcitrant (or refractory) dissolved organic nitrogen* and *soluble non-reactive phosphorus*. Alkalinity is another important parameter to quantify because it can limit the amount of nitrification that can occur, especially if chemical phosphorus removal is used. Another factor to consider is that nonreactive phosphorus is sometimes added to potable water distribution systems in concentrations as high as 10 mg/L to inhibit corrosion, scaling, and/or to reduce lead and copper levels in the distribution system.

Typical WRRF plans tend to assume future load increases; however, it can be much more difficult to operate a BNR facility that is underloaded compared to its ultimate design loading. For example, limitations in the turndown capacity of aeration systems or excessive weir drops at low flowrates may result in undesired carryover of dissolved oxygen to anaerobic or anoxic zones. These issues can be addressed by design and operational modifications; however, the utility must work with the engineer to phase projects with their capital improvement plan and rate model as well as predictions of the regulatory environment and influent loading. Relying on assumptions and "typical" per-capita factors may initially result in overly conservative designs, inefficient operations, and unnecessary spending, whereas investing in wastewater characterization and long-term planning helps ensure success each step of the way.

5.1.1 Climates and Seasons

Utility managers must consider climate and seasonal changes when selecting an appropriate technology and equipping operators with the necessary tools to manage these changes. Many facilities have overcome VFA deficits (common in cold climates) by including fermentation to provide a reliable VFA mixture to drive EBPR and/or denitrification. Fermentation alternatives can be suited to fit essentially any activated sludge configuration, whether it has primary clarifiers or not. Utilities that are not equipped with fermenters can enhance BNR with carbon supplementation; however, chemical selection must be carefully considered. For example, the use of methanol to enhance denitrification requires a higher acclimation period in colder climates and, as such, intermittent dosing (to reduce chemical costs) may not be feasible.

Extremely warm weather can also pose challenges for EBPR because phosphorus accumulating organisms may be outcompeted by glycogen accumulating organisms, which consume VFAs but lack the ability to enhance phosphorus uptake.

5.1.2 Industrial Source Control

Oftentimes, challenges associated with industrial discharges go unnoticed until a utility upgrades the WRRF for BNR because nitrifying bacteria are more sensitive to pH and certain metals and organic compounds than ordinary heterotrophic organisms. In addition to direct dischargers, the CWA also established a regulatory program for indirect dischargers. Pretreatment coordinators, typically at the local utility level and overseen by the state agency (or the U.S. Environmental Protection Agency in nondelegated states), issue and enforce industrial discharge permits to industries within the collection system that meet specific categorical or noncategorical standards (WEF, 2008). By conducting a local limits evaluation (U.S. EPA, 2004), the pretreatment coordinator can determine if inhibitory substances are likely to be present in the wastewater that may result in facility upsets, and the CWA provides legal authority for utilities to identify and enforce reductions in such substances. Water resource recovery facility process upsets can be avoided by having a strong pretreatment program, enforcing against permit violations, and working with industries to select more BNR-friendly chemicals, equalizing their discharges, coordinating batch discharges, and so on.

5.1.3 Collection System Design, Operation, and Maintenance

In addition to sources and climate, influent characteristics can also depend on the design and performance of the collection system. Warm temperatures and long collection system residence times tend to promote hydrolysis and fermentation, which are good for BNR because of their conversion of influent carbonaceous waste into readily biodegradable carbon and VFAs. However, these same conditions can create safety, corrosion, and odor problems in the collection system because of the generation of hydrogen sulfide and other volatile compounds. In some cases, oxidant or nitrate dosing in the collection system may be the best solution to prevent these adverse anaerobic conditions; however, these can be problematic for the downstream WRRF because these solutions also consume biodegradable carbon and VFAs needed to drive BNR processes. Iron salts should be considered as an alternative because dosages for sulfide control typically do not interfere with BNR and may actually enhance phosphorus removal.

Various other chemicals are sometimes used to control root ingrowth; dissolve fats, oils, and grease blockages; or for other collection system maintenance. Products and application practices should be carefully selected to avoid unintended consequences downstream at the WRRF. The bottom line is that problem-free nutrient removal requires the coordinated design and operation of both the collection system and WRRF.

5.2 Wet Weather Flow Management

Generally, it is not technically practical or fiscally responsible to design a BNR facility for peak daily flows that exceed 2 to 3 times the daily average flow, yet many facilities are faced with hydraulic peaking factors greater than these values. While holding basins can be incorporated to the design to help during storm events, it is typically more beneficial to all stakeholders to reduce the amount of I/I through improvements in the collection system. Completely eliminating I/I is generally not feasible; however, by reducing its amount, the cost of conveyance, storage, and treatment can be significantly reduced. Utility managers must also realize that effective I/I reduction typically requires more than just a one-time fix and that its cost effectiveness can vary greatly from one part of a collection system to another and from one system to another. Utility planners should consider the long-term costs and benefits of different I/I reduction measures, along with other alternatives, to arrive at the optimal balance of conveyance, storage, and treatment capacity for a particular situation. *Core Attributes of Effectively Managed Wastewater Collection Systems* (APWA et al., 2010) and *Guide for Municipal Wet Weather Strategies* (WEF, 2013a) should be consulted for further details and guidance for such planning and studies. In many instances, managing additional wet weather flows at the WRRF may be part of the optimal solution.

The inherent nature of biological treatment mechanisms tends to limit their effectiveness beyond relatively steady-state influent characteristics. Utilities with biological treatment processes must always weigh the risks of upset, inhibition, or loss of biomass from wet-weather flows, and these risks tend to be greater for BNR facilities due to the slower growth rates of BNR microorganisms; full recovery from upsets could result in extended periods of substandard treatment performance. In addition to I/I reduction, storage and equalization, *Wet Weather Design and Operation in Water Resource Recovery Facilities* (WEF, 2014) should be consulted for various site-specific operational and design measures to help maximize the wet weather treatment capacity of a particular WRRF. The optimum choice depends on site-specific factors such as influent wet weather flow characteristics, water quality requirements of the receiving stream, the type and condition of existing collection and treatment infrastructure, regulatory policies, and affordability, but typically includes consideration of one or more of the following alternatives:

- Operators may be able to reconfigure their activated sludge process to a "deep step-feed" or contact stabilization arrangement, where a portion of the peak wet weather flows are routed to the end of a plug flow aeration basin or to a separate biocontact basin to help temporarily reduce the solids loading rate to clarifiers and minimize biomass washout;
- Temporarily transferring a significant portion of biomass to offline storage basins can accomplish similar results as the reconfigurations described above. After flows have subsided to acceptable rates, biomass inventory is shifted back to the normal aeration basins;
- Blending with process trains that provide effluent quality equivalent to primary clarification may be a feasible option for many WRRFs. Settling basins, some fine screens, vortex separators, lamella settlers, and "coarse" filtration technologies can generally meet this objective;
- Chemically enhanced settling can provide significantly better effluent quality than conventional primary clarification. Sludge recirculation, ballasted flocculation, and lamella settlers can further enhance clarification performance and decrease footprint requirements;
- High-rate filtration alternatives may offer a chemical-free alternative to chemically enhanced settling options; and
- Fixed-film, integrated fixed-film activated sludge, or ballasted mixed liquor alternatives may offer the potential to treat higher peaking factors than conventional activated sludge systems. However, BNR performance during the peak flow event will tend to decrease somewhat because of the typically cooler and less-concentrated influent characteristics and shorter hydraulic residence times that typically accompany peak wet weather flows. The high velocities of wet weather flows in the collection system also tend to shear collection system biofilms and may also contribute to less fermentation and lower proportions of influent readily biodegradable carbonaceous substrate, such as VFAs. This is another reason to advocate for long-term average permit limits, that is, so that transient, short-term reductions in treatment efficiency that have little to no additional effect on water quality will not result in unnecessary permit enforcement actions.

5.3 Biosolids Management

Successful nutrient removal and recovery requires coordinated design and operation of biosolids facilities with liquid stream process trains. In the case of nutrient recovery, alternatives other than effluent reuse typically require that nutrients be transferred from the WRRF liquid stream to its biosolids stream, which also creates a more nutrient-rich stream, typically increasing the feasibility of nutrient recovery.

Land application of biosolids is an important avenue for nutrient recovery. Increasing phosphorus removal from the liquid stream increases the phosphorus content of the biosolids, which, if left unchecked, might limit land application to phosphorus agronomic rates in some cases (WDNR, 2009). The extent and practical effects of these limitations are highly case-specific and depend on details of both the biosolids management and phosphorus removal practices of the particular WRRF in question. Phosphorus recovery from the biosolids can lower their phosphorus content; however, soil conditions and land use are also factors, and local regulations must also be consulted. The *National Manual of Good Practice for Biosolids* (NBP, 2011) should be consulted for additional details on biosolids management.

Anaerobic digestion is an important biosolids alternative, particularly for energy recovery. It can pose challenges for nutrient removal, but also offers opportunities for nutrient recovery. When biosolids are anaerobically digested, ammonia and phosphate are released and are often recycled back to the liquid stream process in digestate dewatering liquors, which significantly increase nutrient loads to the liquid train and can cause significant operational issues, particularly if recycle load equalization is not provided. Additionally, some utilities have observed decreases in the dewaterability of biosolids and/or increased struvite formation when EBPR and anaerobic digestion have been paired, which can further increase operating costs (Shimp et al., 2014). Metal salts can be used to mitigate these effects; however, they will not lower the phosphorus content of the biosolids nor decrease ammonia recycle loads. As mentioned in Section 4.0, predigestion anaerobic conditioning of WAS can be used to release orthophosphate (Chaparro and Noguera, 2003) as well as magnesium and potassium. Predigestion thickening can shunt these ions around digestion and dewatering and pairs well with struvite recovery to potentially provide multiple benefits besides nutrient recovery. Sidestream deammonification is also rapidly gaining acceptance to cost-effectively further decrease ammonia recycle loads, and sidestream ammonia recovery processes continue to be investigated (Tchobanoglous, 2012).

5.4 Integration of New Processes and Technologies

Increasing nutrient removal and recovery at a WRRF will likely require changing the configuration of existing processes or incorporating processes and technologies unfamiliar to the utility. To mitigate potential risks, operational and maintenance requirements must be well understood and addressed in the design of the new facilities and process control features.

5.4.1 Operations and Maintenance

Operations and maintenance staff will likely need to learn new skills and develop different operation and maintenance procedures to fully implement and optimize nutrient removal and recovery. Biological nutrient removal facilities may require online, real-time nutrient analyzers (ammonia, ammonium, nitrate, nitrite, phosphate, etc.) and other instrumentation (oxidation/reduction potential, dissolved oxygen, etc.) to monitor and optimize conditions in each part of the process. To provide the most benefit, it is important that the instruments be reliable and easy to maintain and that operations staff understand their use in the control of the process. These instruments, in addition to mixers, pumps, blowers and diffusers, require regular preventive maintenance to function properly and efficiently. While significant strides have been made, research and development should continue to further improve instrument reliability and lower maintenance requirements. Maintenance budgets for BNR can be significantly higher than those for carbonaceous-removal only, and utilities often must add additional staff, particularly technicians skilled in maintenance and programming of instruments, process logic controllers, and supervisory control and data acquisition systems.

Regulatory agencies may also require a higher certification level for operators of BNR facilities versus those operating less advanced processes. Operations budgets, therefore, need to account for additional training, professional development hours, and examination and certification fees, in addition to added staff. Other common increases to operations budgets when upgrading to BNR include electrical, chemical, and analytical (laboratory) costs. Additional information regarding the operation of BNR facilities can be found in *Biological Nutrient Removal (BNR) Operation in Wastewater Treatment Plants* (WEF, 2006).

5.4.2 Design of Facilities

From a design standpoint, the level of redundancy needs to be considered. Risk mitigation includes analyzing the configuration of the WRRF before and after nutrient management is in place for reliability, availability, and maintainability to help ensure consistent performance. Tools such as Hazard Analysis and Critical Control Points can help in this regard (U.S. FDA, 2014). Other design and project management tools to assess risk include fault tree analysis and availability analysis, where the utilization of a nutrient unit process can be quantified in terms of annual average operating hours at full capacity versus reduced/derated capacity. Effects from planned and unplanned maintenance can also be assessed.

Risk assessment can be used in the conceptual planning and design phases to determine the number of operating and spared equipment or process trains (e.g., four 33% trains versus three 50% trains). The quantitative tools can also be used to evaluate "n+1" versus "n+2" redundancy levels and establish whether spare units need to be in hot or cold standby, shelf spares, or installed spares.

Specific risks that should to be considered in the design are biological upsets or changes in influent quality that may result in noncompliance. Ways to mitigate these risks might be to include backup chemical feed facilities (metal salts for phosphorus precipitation, a carbon source for denitrification, etc.). The capital funds spent on backup or auxiliary facilities should be weighed against the level of regulatory risk; WRRFs with stringent discharge limits should consider more than one way to accomplish the discharge objectives. Additional information on the design of nutrient removal and recovery facilities can be found elsewhere in this publication as well as in *Nutrient Removal* (WEF, 2010).

5.4.3 Process Modeling

Dynamic models and simulation platforms are being used more frequently to optimize WRRF design and operations (Phillips et al., 2009). These predictive tools can help planners, designers, supervisors, and operators answer "what if" questions about optimal recycle and wasting rates, changes in influent characteristics, biosolids processing effects, or taking basins out of service for maintenance. Successful application of these tools can significantly reduce operating costs, improve effluent quality, and provide powerful training for staff. For the tools to be successfully applied, however, there must be one or more champions within the utility to ensure that they are used and validated regularly with good data. Resources in this regard include *Guidelines for Using Activated Sludge Models* (IWA Task Group, 2013) and *Wastewater Treatment Process Modeling* (WEF, 2013b).

Dynamic WRRF models can also be linked to dynamic collection system and receiving-stream models to provide a comprehensive water quality model, which can be a powerful tool to help evaluate probabilities of various scenarios and alternatives for watershed water quality management.

6.0 INNOVATION RISKS

Nutrient removal and recovery may require the adoption of new or innovative technologies and practices not typical in current WRRFs. Collaborative research and development will help reduce the risks that are inherent to emerging technologies. The Leaders Innovation Forum for Technology is an example of such collaboration (http://www.werf.org/lift), and this publication itself grew out of a collaborative dialogue between stakeholders representing diverse interests (practitioners, consultants, regulators, academia, environmental advocates, and other water sector professionals).

Emerging technologies typically follow an "S-curve" during maturity, with a critical risk point or "chasm" during the initiation phase when they are first piloted in the field. A second chasm risk may also appear early during the adoption phase when technologies are applied at different scales and locations (Abouzeid, 2011). Pilot testing and modeling tools can help mitigate these chasm risks by demonstrating performance on a small scale at a particular location and projecting those results to a larger scale. Public–private partnerships can help offset potential risks and are especially helpful to early adopters as a technology successfully moves past the "pilot chasm" and strives to cross over to the "early-adoption chasm". Long-term trials (6 months to a year) in large-scale pilots or "demonstration-scale" facilities can also be used during this phase to further evaluate scale-up and long-term operational issues and to investigate optimization opportunities before more widespread adoption (PUB, 2014).

To help spur adoption of best management practices, regulatory agencies should explore the concept of "safe harbors" for technology innovators, as discussed in the Johnson Foundation Report (2014) titled, *Crafting A Vision for Nutrient Management in Wastewater Treatment*. Current regulatory schemes often create disincentives by penalizing innovators for minor noncompliance with permits. Regulators could use enforcement discretion or permitting flexibility to allow innovators to operate outside of "normal" permitted limits for short periods of time to explore new technologies that may ultimately prove to significantly enhance nutrient management practices and realize meaningful results.

7.0 INFRASTRUCTURE RISKS

Planning and designs for the collection system and WRRF should keep an eye toward future nutrient management alternatives and include the following considerations:

- Plans for construction at the WRRF should consider leaving space in the facility hydraulic profile to accommodate future processes that may be needed for nutrient management. Consider designing basins that can be efficiently repurposed as nutrient removal and recovery needs change (Scanlan et al., 2008). Additionally, consider aeration design features that may offer greater flexibility and more energy efficiency, such as membrane diffusers for swing zones or cyclic aeration, blower system turndown to prevent excessive aeration, multiple drop zones for tapered aeration, and independent basin mixing versus process air;

- The framework and supporting technologies for source separation and decentralization have matured and may see increasing roles in resource management, particularly in urbanized settings (Larsen et al., 2013). These approaches may require different collection, recovery, and treatment infrastructure for yellow (nutrient-rich), black (carbon-rich), and gray (stormwater and dilute wastes) liquid streams. Source separation may also significantly alter WRRF influent characteristics and performance (Tchobanoglous, 2012); and
- As discussed in Section 5.1.3, fermentation in the collection system can boost BNR performance, but poses potential risks because of accompanying odors and increased corrosion potential. Coordinating the design and operation of a WRRF and collection system is essential for successful BNR.

8.0 REFERENCES

Abouzeid, M. (2011) http://markabouzeid.com/technology-adoption-lifecycle-revisited/ (accessed Feb 2015).

American Public Works Association; American Society of Civil Engineers; National Association of Clean Water Agencies; Water Environment Federation (2010) *Core Attributes of Effectively Managed Wastewater Collection Systems*; National Association of Clean Water Agencies: Washington, D.C.

Ashley, K.; Cordell, D.; Mavinic, D. (2011) A Brief History of Phosphorus: From the Philosopher's Stone to Nutrient Recovery and Reuse. *Chemosphere*, 84 (6), 737–46.

Baeza, J. A.; Guerrero, J.; Guisasola, A. (2014) Can Nitrogen Removal and Enhanced Biological Phosphorus Removal Coexist? *Water Online*, Sept, 18–21.

Benndorf, J.; Chorus, I. (2006) Editorial: Liebig's Minimum Law, the N-P Controversy and the Need for a New Immission Principle in Eutrophication Control—An Extended Editorial. *Acta Hydrochim. Hydrobiol.* 34, 299–303.

Bunch, D.; Kobylinski, E.; Koch, D.; Ratzki, T.; Steichen, M. (2014) What Everyone Should Know About Enhanced Biological Phosphorus Removal. *Water Online*, Sept, 11–16.

Chaparro, S. K.; Noguera, D. R. (2003) Controlling Biosolids Phosphorus Content in Enhanced Biological Phosphorus Removal Reactors. *Water Environ. Res.*, **73** (3), 254–262.

Cubas, F. J.; Novak, J. T.; Godrej, A. N.; Grizzard, T. J. (2014) Effects of Nitrate Input from a Water Reclamation Facility on the Occoquan Reservoir Water Quality. *Water Environ. Res.*, **86** (2), 123–133.

Fitzpatrick, J.; deBarbadillo, C.; Kobylinski, E.; Phillips, H.; Schauer, P.; Shaw, A. (2014) Turning BNR Lemons into Lemonade. *Water Environ. Technol.*, **26** (10), 34–39.

Gabriel, O.; Balla, D.; Kalettka, T.; Schilling, C.; Zessner, M. (2006) Estimating the Influence of Increased Nitrate Loads from Wastewater on Phosphorus Mobility in River Sediments. *Acta Hydrochim. Hydrobiol.*, **34**, 333–345.

International Water Association Task Group on Good Modeling Practice (2013) *Guidelines for Using Activated Sludge Models*; IWA Publishing: London, U.K.

Johnson Foundation (2014) *The Road Toward Smarter Nutrient Management in Municipal Water Treatment;* The Johnson Foundation at Wingspread: Racine, Wisconsin. http://www.johnsonfdn.org/sites/default/files/conferences/whitepapers/14/03/10/cnw_smarter_nutrient_management_0.pdf (accessed Feb 2015).

Lake, B. A.; Coolidge, K. M.; Norton, S. A.; Amirbahman, A. (2007) Factors Contributing to the Internal Loading of Phosphorus from Anoxic Sediments in Six Maine, USA, Lakes. *Sci. Total Environ.*, **373**, 534–541.

Larsen, T. A.; Udert, K. M.; Lienert, J. (2013) *Source Separation and Decentralization for Wastewater Management*; IWA Publishing: London, U.K.

Lindeke, D. R. (2014) Everything I Know About Bio P, the Bugs Taught Me. *Proceedings of the 87th Annual Water Environment Federation Technical Exhibition and Conference* [CD-ROM; New Orleans, Louisiana, Oct 18–22; Water Environment Federation: Alexandria, Virginia.

Melcer, H.; Dold, P.; Jones, R. M.; Bye, C. M.; Takacs, I.; Stensel, H. D.; Wilson, A. W.; Sun, P.; Bury, S. (2003) *Methods for Wastewater Characterization in Activated Sludge Modeling*; Report No. 99-WWF-3; Water Environment Research Foundation: Alexandria, Virginia.

National Biosolids Partnership (2011) *National Manual of Good Practice for Biosolids*; National Biosolids Partnership: Alexandria, Virginia.

Petzoldt, T.; Uhlmann, D. (2006) Nitrogen Emissions into Freshwater Ecosystems: Is There a Need for Nitrate Elimination in all Wastewater Treatment Plants? *Acta Hydrochim. Hydrobiol.,* **34**, 305–324.

Phillips, H. M.; Sahlstedt, K. E.; Frank, K.; Bratby, J.; Brennan, W.; Rogowski, S.; Pier, D.; Anderson, W.; Mulas, M.; Copp, J. B.; Shirodkar, N. (2009) Wastewater Treatment Modeling in Practice: A Collaborative Discussion of the State of the Art. *Water Sci. Technol.,* **59** (4), 695–704.

PUB (2014) *Innovation in Water—Singapore*, 6th ed.; PUB: Singapore.

Scanlan, P.; Klamm, J.; Fitzpatrick, J. (2008) Plant Design for System Flexibility—Minimizing Process Modification Requirements in a Changing Biosolids Environment. *Proceedings of Water Environment Federation Residuals and Biosolids Conference*; Philadelphia, Pennsylvania, Mar 30–Apr 2; Water Environment Federation: Alexandria, Virginia.

Shimp, G. F.; Barnard, J. L.; Bott, C. B. (2014) It's Always Something: Understanding the Connections among Biological Phosphorus Removal, Anaerobic Solids Digestion, Dewatering, and Land Application. *Water Environ. Technol.,* **26** (6), 42–47.

Tchobanoglous, G. (2012) Wastewater Treatment Trends in the 21st Century That Will Impact Water Reuse Projects. *Proceedings of the 3rd International Sede Boqer Conference on Water Technologies*; Ben-Gurion University of the Negev, Israel.

U.S. Environmental Protection Agency (2004) *Local Limits Development Guidance*; EPA-833/R-04-002A; U.S. Environmental Protection Agency, Office of Wastewater Management: Washington, D.C.

U.S. Food and Drug Administration (2014) http://www.fda.gov/Food/GuidanceRegulation/HACCP/ (accessed Feb 2015).

Water Environment Federation (2006) *Biological Nutrient Removal (BNR) Operation in Wastewater Treatment Plants*; WEF Manual of Practice No. 29; McGraw-Hill: New York.

Water Environment Federation (2013a) *Guide for Municipal Wet Weather Strategies*; Water Environment Federation: Alexandria, Virginia.

Water Environment Federation (2013b) *Wastewater Treatment Process Modeling*, 2nd ed.; WEF Manual of Practice No. 31; McGraw-Hill: New York.

Water Environment Federation (2008) *Industrial Wastewater Management, Treatment, and Disposal*, 3rd ed.; WEF Manual of Practice No. FD-3; McGraw-Hill: New York.

Water Environment Federation (2010) *Nutrient Removal*; WEF Manual of Practice No. 34; McGraw-Hill: New York.

Water Environment Federation (2014) *Wet Weather Design and Operation in Water Resource Recovery Facilities*; Water Environment Federation: Alexandria, Virginia.

Water Environment Research Foundation (2014) Full-Plant Deammmonification for Energy Positive Nitrogen Removal. http://www.werf.org/a/ka/Search/ResearchProfile.aspx?ReportId=INFR6R11 (accessed Feb 2015).

Wisconsin Department of Natural Resources (2009) *Introduction to Phosphorus Removal Study Guide*; Wisconsin Department of Natural Resources: Madison, Wisconsin.

7

Case Studies

1.0 INTRODUCTION

Nutrient management examples aligned with the management solutions and roadmap elements outlined throughout this publication are provided in this chapter. Table 7.1 provides a summary of the case studies included in this chapter, along with the corresponding roadmap elements for which they provide examples.

The case studies include those that vary from small community/small water resource recovery (WRRF) facilities (e.g., City of Conrad WRRF, Montana) to large WRRFs (e.g.; Upper Blackstone Pollution Abatement District, Massachusetts) to watersheds (e.g.; Clean Water Services, Oregon), and by no means represent a total and exact summary of all possible nutrient roadmap examples available. Rather, they offer examples and lessons learned from real-life cases where communities and regulators have come together to develop a framework and solutions to the complex nutrient (nitrogen [N] and phosphorus [P]) management solutions facing them.

TABLE 7.1 Select case studies representing key elements of nutrient management.

	Waste load allocations, watershed management	Integration of nutrient recovery and sidestream technologies	Integration with wet weather management	Integration with advanced treatment systems	Integration with natural treatment systems	Water supply reliability and water recycling
City of Conrad WRRF, Montana	✓					
Clean Water Services, Oregon	✓	✓	✓	✓	✓	✓
Marlborough Westerly Wastewater Treatment Facility, Massachusetts	✓			✓		
Upper Blackstone Water Pollution Abatement District WRRF, Massachusetts	✓	✓				
Vadnais Lake, Minnesota					✓	✓
San Francisco Bay, California	✓	✓	✓			✓
Chesapeake Bay Watershed, Virginia, Maryland, Pennsylvania, New York	✓	✓	✓	✓	✓	✓
Broad Run Water Reclamation Plant, Virginia	✓		✓	✓		✓

It is important to note that one of the oldest nutrient programs developed to resolve one of the biggest nutrient challenges of North America is the Chesapeake Bay Program (CBP), and it has many program elements that align with the Nutrient Roadmap solutions and associated lessons learned. Owing to the long history of the program, the solutions developed and used by the publically owned treatment works (POTWs) that discharge to the Bay directly or to its watershed provide an array of technology solutions. However, to meet lower limits such as enhanced nutrient removal and ultra-low levels, in some cases, CBP partners have devised innovative solutions and continue to do so as demonstrated in the leading-edge research into deammonification technologies conducted and implemented by partnering POTWs and academic institutions located in the geographical area.

2.0 CITY OF CONRAD WATER RESOURCE RECOVERY FACILITY

General Information

Name: City of Conrad WRRF
Location: Conrad, Montana
Regulatory Drivers/Requirements: Ammonia, nitrate, total nitrogen, and total phosphorus
Flow: 1.89 ML/d (0.5 mgd) design, 1.32 ML/d (0.35 mgd) actual

FIGURE 7.1 City of Conrad WRRF

Watershed Issues and Solutions

Following the enactment of requirements that include flowrate-based effluent nutrient limits (e.g., 1 mg/L total phosphorus and 10 mg/L total nitrogen for mechanical facilities greater than 1 mgd) along with variance processes for affordability and limits of technology, communities in Montana looked for compliance solutions. Additionally, once these initial limits sunset in 2016, a stepped-down approach to tighter nutrient limits will be in effect. The City of Conrad's facility discharges to an ephemeral stream, with the facility effluent constituting 100% of stream flow at times, and, therefore, has stringent effluent limits. Being a smaller facility located in a smaller community, affordability of complex technology solutions was a concern. The City of Conrad's system is also one of the success stories in the State of Montana, exemplifying the importance of collaboration with regulators.

Institutional Challenges and Solutions

Challenges. The City faced a number of challenges as part of conversion to nutrient removal that included a recent change from aerated lagoon to activated sludge treatment; bitterly cold winters; solo operator; facility designed for ammonia removal, not for total nitrogen or total phosphorus removal; regulatory requirement to use all equipment as designed.

Solutions. Montana Department of Environmental Quality (DEQ) is working with the facility owners and staff, and has instituted targeted education and on-site technical assistance programs. Montana DEQ is supportive of alternative operations and maintenance (O&M) practices.

Technology Solutions

Technology. Conversion from aerated lagoon to activated sludge for ammonia removal; aeration tank covered to reduce freezing issues. Also, aeration tank was covered with a floating insulated cover for temperature control. Air is cycled to create alternating aerobic and anoxic conditions for total nitrogen removal. Sludge storage tank is operated without aeration and the waste activated sludge is recycled to the aeration tank for achieving biophosphorus.

Operations. Process control changes, instead of new technologies, are providing for total nitrogen and total phosphorus removal. Operator education and familiarity with the process elements supports success.

Lessons Learned and Pros/Cons of Approaches Taken

Operations first. Notwithstanding conventional wisdom to the contrary, the day-to-day operations of many (perhaps most) WRRFs can be changed to provide some level of nutrient removal. By experimenting and optimizing operations, more sustainable nutrient removal can be achieved while providing capital and O&M savings.

Regulatory flexibility/support. A supportive regulatory mindset is invaluable. Montana DEQ's sponsorship of classroom training and on-site technical support empowered WRRF personnel to experiment with process changes. Instead of insisting that all equipment be used as designed, Montana DEQ supported and even guided staff in straying from the O&M manual. Montana DEQ and the facility owners believe that, through empowerment of the operators with this knowledge and success, achieving further nutrient reduction down the road will be possible.

An emphasis on habitat. By focusing on the development and maintenance of ideal biological habitats for nitrification, denitrification, volatile fatty acid uptake by PAOs, and orthophosphorus assimilation by PAOs, facility personnel found opportunities for nutrient removal without large expenditures.

Energy savings. Facilities that aerate sufficiently to remove ammonia can typically complete nitrogen removal by creating the low-oxygen habitats for nitrate removal while reducing electrical consumption.

Reduced sludge production. The creation of strong aerobic and strong anoxic zones not only provides conditions that serve as a "selector" to control filamentous bacteria, there is an increase in endogenous respiration, which results in reduced sludge generation.

3.0 CLEAN WATER SERVICES

General Information
Name: Clean Water Services (CWS)
Location: Hillsboro, Oregon
Regulatory Drivers/Requirements: Tiered ammonia limit and total phosphorus limit of 0.1 mg/L, temperature
Flow: Combined dry weather discharge flow greater than 435 ML/d (115 mgd)

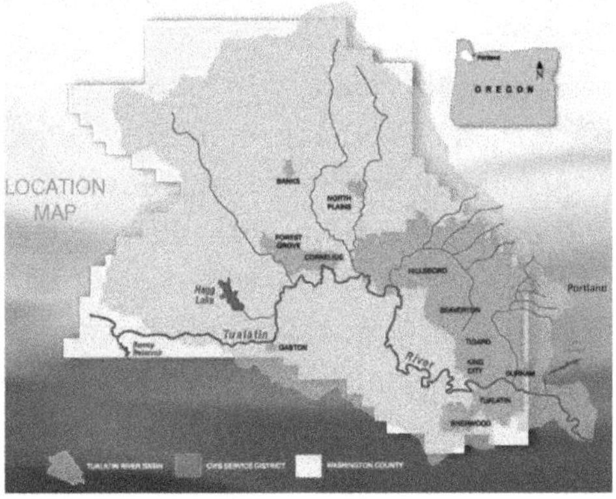

FIGURE 7.2 Tualatin River watershed.

Watershed Issues and Solutions

The listing of many of Oregon rivers, including the Tualatin River, as habitat for endangered cold water fisheries has resulted in temperature total maximum daily loads (TMDLs) on the listed rivers. Located in a historically agricultural area, parts of which are increasingly urbanized, the Tualatin River watershed drains 1844 km^2 (712 sq miles), and has low nutrient limits. The National Pollutant Discharge Elimination System (NPDES) permit is a watershed-wide permit that allows trading of effluent nutrient loads between the facilities.

Clean Water Services District (hereafter referred to as the "District"), which operates an intricate system of mechanical and natural treatment system facilities to manage the loads discharged to the watershed, also has implemented an innovative temperature management plan that includes an extensive shading program in the basin. More than a few million trees have been planted along tributaries and the main stem of the river.

Institutional Challenges and Solutions

The District is a water resources management utility committed to protecting the water resources in the Tualatin River Watershed, located in a western suburban area of Portland, Oregon. The District serves more than 551,000 customers, providing healthy rivers and streams through innovative wastewater and stormwater services, flood management projects, water quality and stream enhancement projects, fish habitat protection, and more. The District manages four WRRFs located along the Tualatin River. Flexibility is provided by force mains connecting the Forest Grove, Hillsboro, and Rock Creek facilities.

The main challenge experienced is the need to continue to respond to the watershed needs as the nutrient trading program and discharge temperature requirements were developed and finalized. Through integrated planning and communication with the regulatory agencies, an equitable and effective trading program was developed to protect the Tualatin River, which is used as a regional drinking water supply and for agricultural irrigation and recreational uses.

Technology Solutions

The NPDES permit includes a tiered ammonia limit based on river flow. Nitrification is required in the dry season and extends into the winter, depending on river flow. Phosphorus limit is nominally 0.1 mg/L, with a load limit based on monthly average values.

Forest Grove Facility. This facility treats an average of 95 ML/d (25 mgd) of wastewater during wet weather only, with dry weather treatment currently provided at the Rock Creek Facility. Future plans include dry weather treatment biological phosphorus removal at the facility and nitrification for a portion of the flow provided in the adjacent Fernhill Wetlands.

Hillsboro Facility. This facility provides secondary treatment of up to 113 ML/d (30 mgd) during the wet weather season only. Dry weather treatment is currently provided at the Rock Creek Facility. Future plans include dry weather treatment biological phosphorus removal at the facility and nitrification for a portion of the flow provided in the Fernhill Wetlands adjacent to the Forest Grove Facility.

Rock Creek Facility. This facility treats an average of 132 ML/d (35 mgd) of wastewater, with a greater secondary treatment capacity for wet weather treatment only. Dry weather treatment includes nitrification and biological phosphorus removal. The facility includes phosphorus recovery using WASSTRIP and OSTARA.

Durham Facility. The facility treats an average of 98 ML/d (26 mgd) of wastewater, with a secondary treatment capacity of up to 195 mgd during the wet weather season. Dry weather treatment includes nitrification and biological phosphorus removal. The facility includes phosphorus recovery using WASSTRIP and OSTARA, producing up to 454 tonnes (500 tons) of Crystal Green® fertilizer per year.

Fernhill Wetlands. Fernhill Wetlands is part of more than 283 ha (700 ac) in Forest Grove owned by CWS and managed in partnership with the City of Forest Grove and the Fernhill Wetlands Council. The wetlands system is a Natural Treatment System and creates an ecological bridge between treatment and the watershed, where water is cleansed further, cooled, and naturalized before its return to the river.

4.0 MARLBOROUGH WESTERLY WASTEWATER TREATMENT FACILITY

General Information
Name: Marlborough Westerly Wastewater Treatment Facility
Location: Marlborough, Massachusetts
Regulatory Drivers/Requirements: Effluent dominated, impounded river; total Phosphorus limit of 0.1 mg/L (April–October) and 1.0 mg/L (November–March); ammonia limit of 2 mg/L (June–October) and 10 mg/L (November–May).
Flow: Design average annual flow, 15.6 ML/d (4.15 mgd); maximum daily, 28.4 ML/d (7.5 mgd); peak hourly, 43.8 ML/d (11.6 mgd)

FIGURE 7.3 Marlborough Westerly Wastewater Treatment Facility.

Watershed Issues and Solutions

In 1999, the Massachusetts Department of Environmental Protection (MassDEP) wrote to the communities along the Assabet River in Massachusetts looking for them to start planning to meet the highest and best practicable treatment for phosphorus removal at the four POTWs that discharged to the effluent-dominated river. Compounding the issue were the seven impoundments located along Assabet River that created an environment that promoted eutrophic conditions in the river. In 2005, the U.S. Environmental Protection Agency (U.S. EPA) and MassDEP issued NPDES permits for the four facilities, requiring them to meet an average monthly total phosphorus limit of 0.1 mg/L (April to October) and 1.0 mg/L (November to March). In addition, a cold weather (November to May) ammonia limit of 10 mg/L was added to the permit to supplement the 2-mg/L ammonia limit from June to October.

Institutional Challenges and Solutions

At the time the permit was issued, few treatment technologies would guarantee a total phosphorus limit of 0.1 mg/L, and issues related to nonreactive phosphorus were emerging. At higher permit limits, nonreactive phosphorus is less of a concern, but, as limits are reduced, nonreactive phosphorus can represent 30% or more of the effluent total phosphorus and can limit a facility's ability to meet the extremely low limits. The TMDL that was completed for the Assabet River acknowledged that sediment flux, nonpoint sources, and background conditions in the river were responsible for more than 25% of the phosphorus load to the river. With the reductions from the four POTWs, it was estimated that phosphorus load to the river would be reduce by 78%.

Technology Solutions

As a part of the planning and preliminary design process, a pilot study was performed using three promising technologies: CoMag™ by Cambridge Water Technologies (now Evoqua), Actiflo® by Kruger, and BluePro® by BlueWater Technologies. Subsequent to piloting, a request for proposal was prepared to assess alternatives based on cost and noncost issues. As a result, BlueWater Technologies was selected as the preferred vendor for the Marlborough Westerly Wastewater Treatment Facility. Design and construction was completed under an Administrative Consent Order. The facility has been operational since January 2012 and is achieving exceptional effluent quality (typically < 2 mg/L biochemical oxygen demand and total suspended solids, < 0.1 mg/L total phosphorus, and < 0.06 mg/L ammonia). As part of the upgrade, a new headworks was installed, mechanical aeration was converted to fine-bubble diffused air with high-efficiency turbo blowers, an intermediate pump station was required to pump to the BluePro® system, and UV disinfection replaced gaseous chlorine. New chemical feed systems were also included.

Lessons Learned and Pros/Cons of Approaches Taken

Based on 2012 data, since the four POTWs became operational, the seasonal phosphorus load to the Assabet River have been reduced by 94%, as compared to 1999 data. The total daily phosphorus load from POTWs to the river represents one-half of the requirements of the TMDL, as presented. The total construction cost to achieve these limits at the four POTWs (including baseline improvements to the facility to upgrade other systems) was $100.5 million, for a total design flow of 61.6 ML/d (16.3 mgd). On a unit-cost basis, the improvements ranged from $1.3 to 1.75/L ($5 to 6.6/gal) treated (based on average design flow). Total annual O&M costs increased from approximately $50,000 to $140,000 per mil. gal treated.

Although located in a water-rich part of the country, effluent from this treatment facility could be considered for re-use. River sampling conducted by the local watershed group has revealed that, in 2013, median total phosphorus concentrations in the Assabet River main stem were, for the first time since monitoring commenced, < 0.05 mg/L (Figure 7.4). Their report indicates that more long-term data are necessary to determine if eutrophic conditions of impounded sections of the river have improved with the reduction in total phosphorus discharged from the POTWs.

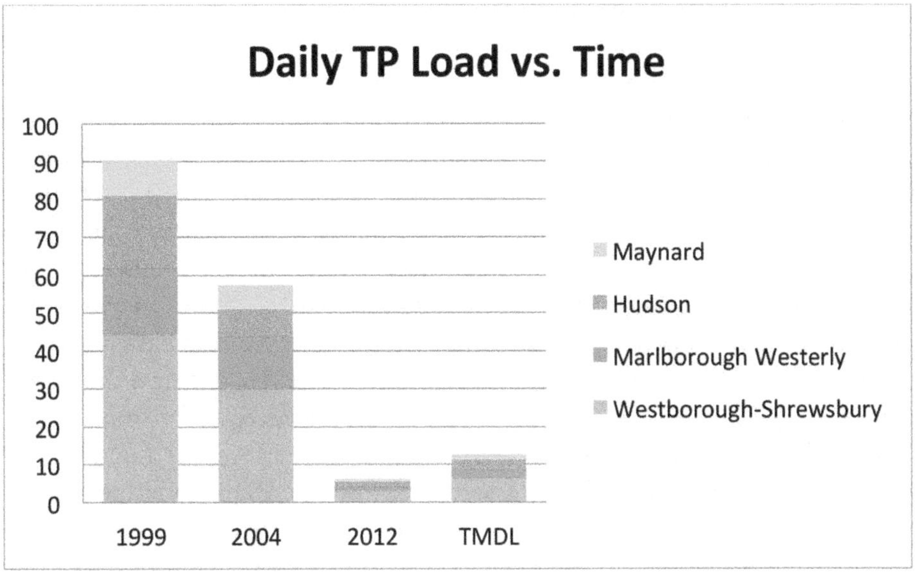

FIGURE 7.4 Reductions in daily total phosphorus loads to the Assabet River.

5.0 UPPER BLACKSTONE WATER POLLUTION ABATEMENT DISTRICT

General Information
Name: Upper Blackstone Water Pollution Abatement District (UBWPAD)
Location: Millbury, Massachusetts
Regulatory Drivers/Requirements: Nutrient reduction, seasonal limits of 5 mg/L total nitrogen (May–October); 0.1 mg/L total phosphorus (April–October)
Flow: Design average annual flow, 170 ML/d (45 mgd); maximum daily, 303 ML/d (80 mgd); peak hourly, 605 ML/d (160 mgd).

FIGURE 7.5 The UBWPAD facility.

Watershed Issues and Solutions

The UBWPAD owns and operates a WWRRF in Millbury, Massachusetts, that treats wastewater from domestic, commercial, and industrial sources from the City of Worcester (the second largest city in New England) and surrounding communities. The average annual design flow for the treatment system is 170 ML/d (45 mgd), with a peak hourly flow of up to 605 ML/d (160 mgd) because a portion of the service area is served by a combined sewer system. The advanced treatment system is designed for a peak hourly flow of 454 ML/d (120 mgd). The facility discharges to the headwaters of the Blackstone River, travels about 72 km (45 miles), and drops 137 m (450 ft) in elevation to Narragansett Bay in Rhode Island. The Blackstone River, once considered the hardest working river in the nation, was also the first river in the nation to use hydropower to spawn the industrial revolution. Nineteen impoundments still exist on the river. The river system shows signs of stress, which is common in urban waterways, and sediment reflects the historic industrial uses. The impoundments create an environment that encourages eutrophication in the river behind the dams, and Narragansett Bay in Rhode Island has been affected by high nitrogen loading and hypoxic conditions. In 2001, U.S. EPA issued an NPDES permit requiring UBWPAD to report total nitrogen and to achieve 0.75 mg/L total phosphorus seasonally. In 2008, U.S. EPA revised these limits to 5 mg/L total nitrogen and 0.1 mg/L total phosphorus in warm weather and 1.0 mg/L in the winter months.

Institutional Challenges and Solutions

Having spent $180 million on treatment facility improvements in recent years to meet the 2001 permit limits, UBWPAD, faced with the 2008 limits, chose to appeal the permit. After a number of years in the court system, UBWPAD was unsuccessful in its appeal. The upgraded advanced treatment system went online in late 2009 (to meet the 2001 limits) and UBWPAD has been achieving effluent quality better than expected (although it cannot achieve the 2008 limits). The UBWPAD continues to implement measures to improve operations to get them closer to meeting the 2008 limits, and is piloting innovative technology for sustainable nutrient recovery.

Technology Solutions

The design used to meet the 2001 limits included an enhanced biological nutrient removal (EBNR) system using an anaerobic–anoxic–oxic (A2/O) system (designed with the flexibility to operate in an A/O mode or an Modified Ludzack–Ettinger mode. The design also retained the ability to operate in a step feed mode. Each of the four aeration tanks are divided into seven compartmentalized tanks to create the anaerobic, anoxic, and aerobic zones. For the A2/O process, the tanks consist of a three-stage anaerobic "selector zone" (A, B, and C), followed by a two-stage anoxic zone (D and E), and, finally, an aerobic zone (F and G). Internal recycle pumps allow nitrified mixed liquor recycle from zone G to zone D for denitrification in the A2/O mode. An alkalinity addition system was originally designed to deliver sodium hydroxide to the aeration tanks, but has been modified to feed magnesium hydroxide. Since the upgraded facility went online in late 2009, UBWPAD has been able to achieve a total nitrogen limit of < 6 mg/L year and a total phosphorus limit of < 0.4 mg/L total phosphorus on an annual average basis without the addition of methanol or supplemental carbon and without effluent filtration.

Lessons Learned and Pros/Cons of Approaches Taken

Achieving biological nutrient removal on a combined sewer system operated under cold weather conditions and with limited alkalinity can be challenging. However, a well-run EBPR system can achieve lower effluent total nitrogen and total phosphorus limits than previously expected. Biological phosphorus removal performs better in the colder months. Effluent quality can be enhanced by properly managing pH and solids retention time, but there is a delicate balance between nitrogen and phosphorus removal. Real-time phosphate and nitrogen analyzers help operators control the EBNR system. Step feed can help effluent performance under high-flow conditions.

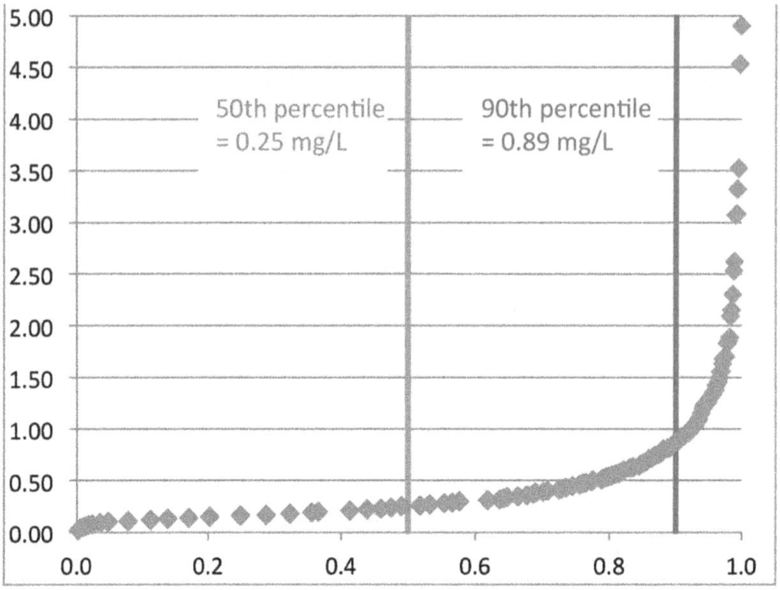

FIGURE 7.6 Final effluent total phosphorus concentrations (January 2013 through January 2014).

Targeted chemical addition (i.e., metal salts for phosphorus removal) and supplemental carbon for nitrogen removal, may improve performance (piloting ongoing). As with any biological system, the system has its ups and downs. Promoting stochastic permitting for nutrients (seasonal or annual median, rather than monthly averages on a mass loading basis) can be appropriate, as illustrated in Figure 7.6.

6.0 VADNAIS LAKE HYPOLIMNETIC OXYGENATION FOR PHOSPHORUS MANAGEMENT

General Information
Name: Vadnais Lake Hypolimnetic Oxygenation for Phosphorus Management
Location: Ramsey County, Minnesota
Regulatory Drivers/Requirements: Taste and odor for drinking water supply
Flow: 163-ML/d (43-mgd) average daily, 476-ML/d (126-mgd) maximum flow

FIGURE 7.7 Pure oxygen linear diffuser system on Vadnais Lake.

Watershed Issues and Solutions

Saint Paul Regional Water Supply pumps water from the Mississippi River into a chain of lakes to provide raw water for the McCarrons Water Treatment Plant. Phosphorus loading from the Mississippi River and local watershed has impaired water quality. Hypolimnetic aeration and ferric chloride injection (1989 to 2010) sharply reduced internal phosphorus loading to Vadnais Lake, which supplies the McCarrons Water Treatment Plant. Ferric iron injection treatment of Mississippi River and local watershed flows significantly reduced external phosphorus loading. Hypolimnetic hypoxia persisted, however, allowing internal phosphorus loading that stimulated algae growth responsible for raw water taste and odor problems.

Institutional Challenges and Solutions

Saint Paul Regional Water Services has conducted systematic improvements to raw water quality and water treatment since the late 1980s. This approach has yielded improvements to water quality in the two principal reservoirs, Pleasant Lake and Vadnais Lake; local watershed improvements; and dramatic improvement in drinking water customer satisfaction. In this approach, the reservoirs and the watershed are integral to overall water treatment.

Technology Solutions

Hypolimnetic aeration demonstrably improved water quality in Pleasant Lake and Vadnais Lake. Replacement of this aging infrastructure with pure oxygen linear diffuser systems (2011 to 2013) further improved water quality by eliminating hypolimnetic anoxia (Figure 7.8) and significantly reducing internal phosphorus loading (Figure 7.9). The apparent hypolimnetic oxygen demand dropped from 2.06 $g/m^2 \cdot d$ in 2010 to 0.22 $g/m^2 \cdot d$ in 2014, which represents a substantial remediation of accumulated sediment oxygen demand.

Lessons Learned and Pros/Cons of Approaches Taken

Hypolimnetic aeration proved to be a far more effective remediation strategy than hypolimnetic aeration. The Vadnais Lake basin is just deep enough (15 m) for the hypolimnion to withstand the mixing energy of linear diffusers without destratifying. Shallower reservoirs may require other means of oxygenation to control internal nutrient loading. Viewed historically, inbasin remediation strategies have been dramatically successful at reducing internal phosphorus loading, reducing the median hypolimnetic total phosphorus by 95% over three successive stages with technical improvement. Where control of external phosphorus loading is impracticable or insufficient to

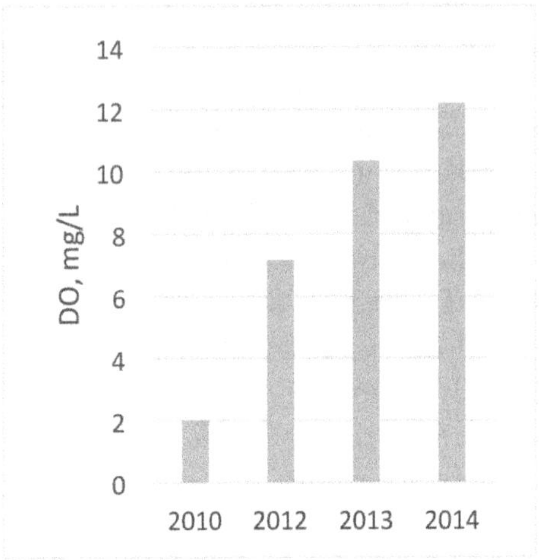

FIGURE 7.8 Mean hypolimnion dissolved oxygen in Vadnais Lake (2010 was the last year of hypolimnetic aeration, and 2012 was the first year of hypolimnetic oxygenation).

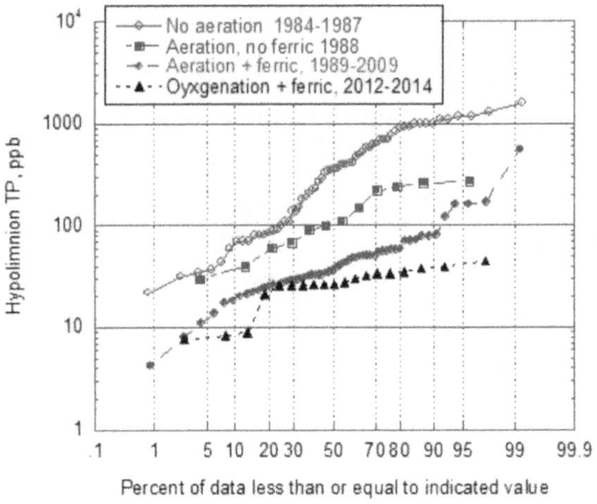

FIGURE 7.9 Hypolimnion total phosphorus historic comparison.

meet water quality goals, oxygenation of the hypolimnion is a viable nutrient control strategy for impaired lakes and reservoirs.

7.0 BROAD RUN WATER RECLAMATION PLANT

General Information
Name: Broad Run Water Reclamation Plant
Location: Loudoun County, Virginia
Regulatory Drivers/Requirements: Nutrient loads management, water reclamation
Flow: 45-ML/d (12-mgd) average flow

FIGURE 7.10 Broad Run Water Reclamation Plant.

Watershed Issues and Solutions

The Chesapeake Bay watershed spans more than 166 000 km^2 (64 000 sq miles). It encompasses parts of six states (i.e., Delaware, Maryland, New York, Pennsylvania, Virginia, and West Virginia) and the entire District of Columbia. Broad Run Water Reclamation Facility (BRWRF) is a state-of-the-art facility that treats wastewater to the limits of technology for the protection of the Capital region's drinking water and is one of the 483 WRRFs discharging to the Chesapeake Bay. The facility is subject to the Dulles Watershed Regulation that requires stringent standards because it is located upstream of a drinking water supply in the Potomac River. The facility is also subject to other specific state regulations as well as Chesapeake Bay regulations. These stringent discharge limits include a total nitrogen limit of 4 mg/L, a total phosphorus limit of 0.1 mg/L, a total Kjeldahl nitrogen limit of 1.0 mg/L, and a chemical oxygen demand (COD) limit of 10 mg/L. The BRWRF sets a new worldwide technology standard for environmental protection and water reclamation.

Technology Solutions

The 12-mgd Broad Run Water Reclamation Plant, located in the Chesapeake Bay watershed, is owned and operated by Loudoun Water in Ashburn, Virginia, and is designed to achieve an average effluent total phosphorus of 0.1 mg P/L by simultaneous biological and chemical phosphorus removal and an average effluent total nitrogen of 4 mg N/L. Water reclamation to near potable water quality is also required, including an effluent COD of 10 mg/L. The facility consists of preliminary treatment, primary treatment, flow equalization, fine screening (2 mm), nutrient removal membrane bioreactor (MBR), activated carbon, and UV disinfection. Waste activated sludge is centrifuge-thickened, combined with primary sludge for stabilization in anaerobic digesters, and centrifuge-dewatered before land application. The design of the MBR uses a five-stage biological process, with the capability to add external carbon to the second anoxic zone and alum to the bioreactor effluent provided. The facility has been operational since April 2008.

The facility can be operated to maximize biological nitrogen removal by operating at a nitrified mixed liquor recycle rate 7 times the influent flow and with a larger first anoxic zone. Additionally, an automatic dissolved oxygen control system was incorporated to allow the process to function properly and maintain desired aerobic zone dissolved oxygen concentrations. This allowed the performance of the nutrient removal system to meet and exceed the design goals. Typically, external carbon addition is not needed.

Additionally, the facility incorporates a fully automated facility-wide control system to optimize treatment, energy, and chemical consumption.

Lessons Learned and Pros/Cons of Approaches Taken

Combining very stringent nitrogen and phosphorus limits with other treatment requirements such as effluent COD and cold temperature operation can be challenging. These challenges are successfully overcome through design features that allow management of biology and process features (e.g., mixed liquor recycle rates, solids retention times, and chemical polishing) and advanced technology such as membranes that provide a barrier for solids/liquid separation, resulting in superior effluent quality. Pilot testing to refine the key design elements was conducted with the participation of facility staff, which also allowed them to learn more about the technology and nutrient removal ahead of the facility startup. Total nitrogen concentrations averaged less than 3.2 mg/L for the first year and thereafter without supplemental methanol addition.

8.0 SAN FRANCISCO BAY

General Information
Name: San Francisco Bay
Location: California
Regulatory Drivers/Requirements: Bay water quality, nitrogen, phosphorus

FIGURE 7.11 San Francisco Bay nutrient loads.

Watershed Issues and Solutions

The San Francisco Bay is the largest estuary on the Pacific Coast. Its value to the California economy is enormous, and its biodiversity significantly contributes to the quality of life for nearly 7 million residents of the Bay Area. According to the San Francisco Regional Water Quality Control Board (RWQCB), the Bay has not experienced nutrient-related environmental problems, despite being recognized as nutrient-enriched, as it has long been considered resilient to the effects of nutrients because of strong tidal mixing in the Bay; high bay turbidity, which limits light penetration; and high filtration by clams. The levels of phytoplankton and microalgae that form an essential component of the aquatic food chain is, therefore, lower than would be expected in a nutrient-enriched system. Recent studies have shown that the Bay's historic resilience to nutrient loading may be weakening. As a result, nutrients are a growing concern for the health of the Bay ecosystem.

Institutional Challenges and Solutions

Numerous scientific studies are being conducted by several entities to understand the effect of nutrients on the San Francisco Bay. While these studies are ongoing, the Bay Area Water Resource Recovery Facilities, known collectively as the Bay Area Clean Water Agency, is working with regulators to ensure that future nutrient effluent limits for WRRFs are based on sound science.

The first regulatory step taken was the issuance of a Regional Watershed Permit (order no. R2-20140-0014). This permit includes four key elements that provide a systematic approach to evaluating nutrient load reduction opportunities at WRRFs (if supported by sound science):

- Facility optimization;
- Sidestream treatment;
- Facility upgrades; and
- By other means (includes "outside-the-fence" opportunities, such as recycled water).

The permit allows WRRFs to perform the permit tasks collectively as a group or by themselves. All 37 participating WRRFs decided to perform the efforts collectively as a group. The Regional Watershed Permit efforts began in 2014 and will continue through 2018.

8

Conclusions

The wastewater treatment industry currently faces a significant challenge. While many utilities are struggling to achieve adequate removal of organic and solid pollutants and nutrients, such as nitrogen and phosphorus, to achieve regulatory mandates, society is recognizing the need to conserve energy and recover nutrients, which are in scarce supply or are causing widespread primary and secondary environmental damage. This implies a collective change in thinking from a mindset of nutrient removal to one of resource recovery and reuse. Because wastewater treatment plants (WWTPs) are expensive, long-term investments, utilities are reluctant to abandon facilities that no longer serve these new societal goals. Reluctance to abandon "stranded assets" is, therefore, a significant factor in making decisions about future investments. As a result, utilities are faced with the challenge of charting a course that meets current needs for regulatory compliance while looking toward a future where critical nutrients are not just removed, but recovered for reuse.

The most well-developed strategy for nitrogen recovery is to recycle wastewater treatment biosolids to productive agricultural or forestry use. Land application, however, only recovers and reuses about 15 to 25% of the nitrogen influent to the facility. Some processes for recovery of phosphorus by precipitation from anaerobic recycle streams recover small amounts of nitrogen as well. However, typical strategies for nitrogen removal from the liquid treatment scheme discharge nitrogen to the atmosphere, thus thwarting recovery. Strategies for fuller recovery of nitrogen from the main liquids flow stream of the facility are only in the developmental stage, but should be further pursued.

Recovery of phosphorus is better developed than for nitrogen. Liquid stream processes using enhanced biological phosphorus removal can concentrate liquid stream phosphorus into biological biosolids streams from which it can be recovered and reused. Return streams from anaerobic biosolids treatment processes can recover phosphorus by precipitation into struvite, for example. Wastewater contains other minerals that are currently mined, such as sulfur, boron, magnesium, and calcium. Presently, the best way to recover these minerals is by reuse of wastewater treatment biosolids.

The change in attitude from removal to recovery is highlighted in WEF's recent recommendation that utilities begin using the term, *water resource recovery facility* (WRRF), in place of *WWTP*. The transition from nutrient removal to nutrient recovery will not take place unless and until utilities embrace the need for it and incorporate it to their staff training, facilities' planning, and facility upgrade programs.

Specific conclusions from *The Nutrient Roadmap* include the following:

- Chapter 1—"Introduction"
  - The purpose of the Nutrient Roadmap is to promote a transition from wastewater treatment to resource recovery;
  - Drivers for this change are population growth, environmental concern, and water quality protection; and
  - Planning for wastewater management must consider future trends and regulatory requirements.
- Chapter 2—"Setting the Stage for Nutrient Management"
  - Develop a vision,
  - Train and empower staff,
  - Develop financing strategies,
  - Plan and prioritize for compliance,
  - Establish goals, and
  - Become climate ready.
- Chapter 3—"Understanding the Environment"
  - By developing an informed decision-making process, the utility should then be in a position to select a pathway for nutrient recovery or removal, and
  - Consider factors both inside and outside the fence of the WRRF and develop solutions that can achieve resource recovery system-wide.
- Chapter 4—"Resource Recovery"
  - There are opportunities for nutrient recovery from the treatment process in the liquid stream for irrigation, in the biosolids for land application, and as a manufactured nutrient product (fertilizer);
  - Utilities need to assess treatment and resource recovery technologies holistically by considering the nutrient cycle, the water cycle,

and energy use (in other words, to fully understand the water–nutrient–energy nexus in their service area);

  - In areas where stringent nutrient standards may drive treatment process selection, reclamation should be considered; and
  - Utility owners need to understand their service area, know the large water users in the system, and consider means to provide an alternative water source to these users, which may include agriculture, industry, power plants, and irrigation.
- Chapter 5—"Alternatives Evaluation"
  - A systematic process of alternatives evaluation should be used to identify opportunities and optimum strategies for nutrient removal and recovery,
  - Process comparisons should start with the entire universe of possibilities and proceed to detailed cost-and-effect evaluation through to final selection, and
  - A traditional "plan–do–check–adjust" approach can be used as a general basis for an optimization program.
- Chapter 6—"Identifying and Managing Risks"
  - Successful nutrient management requires more holistic and adaptive regulatory approaches than those traditionally used to control toxicants and conventional pollutants;
  - Nutrient discharge requirements should be based on local receiving water needs and fate and transport through the ecosystem;
  - Release of nutrients from benthic sediments should be considered when developing nitrogen and phosphorus discharge limits. In some cases, nitrates can actually be beneficial to water quality;
  - Financial risks of nutrient removal are similar to those of conventional facilities;
  - Optimum design and operation must consider diurnal and seasonal variations in waste characteristics from domestic and industrial sources and collection system operations and maintenance practices;
  - Nutrient management facilities may be more sensitive to wet weather flow upsets than conventional carbonaceous processes and, therefore, require a more careful weighting of upset risk;
  - Anaerobic digestion is an important biosolids alternative, particularly for energy recovery. It can pose challenges for nutrient removal, but also offers opportunities for nutrient recovery;
  - Risk mitigation includes analyzing the configuration of the WRRF before and after nutrient management is in place for reliability,

availability, and maintainability to help ensure consistent performance; and
  - Nutrient removal and recovery may require adoption of new or innovative technologies and practices not typical in current WRRFs.
- Chapter 7—"Case Studies"
  - Eight examples of nutrient management programs illustrating a range of techniques and strategies from Chesapeake Bay watershed planning to individual treatment facilities are presented;
  - System-wide nutrient management solutions have proven effective at providing broad environmental benefits in addition to reduction of nutrient loads;
  - Techniques including biological, chemical, and physical nitrogen and phosphorus removal and recovery are presented;
  - Innovative permitting strategies, including annual permit and load-sharing regional permits, are discussed; and
  - The least well-developed area of research for nutrient recovery is for nitrogen recovery from liquid treatment streams. This needs to be emphasized in future.

Appendix

TOPIC 1 Supporting Nutrient Management

	Plan	Prepare & Implement	Evaluate & Improve
Nutrient Vision	Develop Vision • Evaluate goals and current drivers ○ Regulatory requirements ○ Watershed water quality and total maximum daily loads ○ Effects of nutrient management beyond the watershed (long distance aquatic connections, atmospheric) • Establish aspirational goals and sub-goals ○ Nitrogen and phosphorus recovery ○ Energy and water recovery • Become climate-ready ○ Reduce direct and indirect emissions	*Communicate Internally* • Determine level of treatment required to meet goals • WRRF leadership/managers link the nutrient management vision to staff performance plans • WRRF leadership/managers incorporate nutrient goals and key performance indicators into strategic plan	*Communicate Externally* • WRRF shares nutrient vision with external stakeholders and the sector • Plans are in place to embrace external market changes • Review performance against goals • Reassess long-term goals
Staff Development and Alignment	*Set Training Plan* • Staff performance plans include related training and professional development to support WRRF's vision for nutrient management • WRRF fulfills training needs for all relevant positions: management, engineering, and operations	*Train and Support Staff* • WRRF staff are trained in operations • WRRF staff maintain/increase knowledge of emerging technologies through information sharing events	*Empower Staff* • WRRF leadership/managers establish incentives for nutrient results • WRRF leadership/managers empower staff to make changes toward nutrient goals • WRRFs mentor and guide other local and regional utilities to advance nutrient goals
Financial Viability	*Identify Funding Options* • Identify economic incentives and costs for nutrient removal and/or recovery • Identify size of WRRF where recovery is cost-effective (Am I big enough or too big?) • Develop financial strategy to support nutrient projects • Consider alternate financing methods ○ Private vs public finance and operations ○ Alternative project delivery (design-build-operate, build-own-operate-transfer, etc.) ○ Sale of greenhouse gas credits for N_2O-emission reduction and renewable energy generation	*Budget for Success* • Use lifecycle analysis for project decision-making • Nutrient management/recovery is considered on all capital project designs, in operating budget decisions, and standard operating practices	*Invest in the Future* • WRRF's recovered resource revenues generate sufficient funding to invest in other priorities and reduce upward pressure on rates

TOPIC 2 Understanding the Environment

	Plan	Prepare & Implement	Evaluate & Improve
Data and Characterization	Collect Information • Parameter monitoring for decision-making (nitrogen species, carbon, greenhouse gas emissions, etc.), potentially includes: ○ Upstream • Collection system • Industrial dischargers ○ At facility • Influent • In-facility • Effluent ○ Environment • Receiving body • WRRF considers frequency of sampling	Analyze Data • Process information to understand options, examples include: ○ Carbon to nitrogen ratio ○ Carbon to phosphorus ratio • Assess current facility performance for nutrient removal • Understand current removal of nitrogen species (incidental/intentional) • Assess receiving-water impacts/needs ○ Water quality degradation because of nutrient discharges ○ Understand greenhouse gas emissions	Proactively Use Data • Use data to improve: ○ Facility operations ○ Trading programs ○ Source control ○ Next design upgrade
Modeling	Select Model • Ensure that data gathered is useful for modeling efforts	Modeling for Design • Use water quality modeling to identify/quantify receiving water response to nutrients ○ How do nutrient effluent concentrations change and affect receiving waters? ○ Decision making support for appropriate technology for various performance levels • Model the treatment process to understand its constraints and opportunities ○ Redundancy in design • Determine whether nitrogen removal technology will require major capital investment • Understand effects on greenhouse gas emissions	Modeling for Operations • Utilize molecular tools • Use modeling in decision-making ○ Determine carbon tradeoffs with nitrogen removal ○ Nitrogen removal requires more supplemental carbon if using bio-P ○ Max denitrifying PAOs • Develop digester phosphorus precipitation models focusing on critical chemistry • Understand settling characteristics – must not compromise settleability or dewaterability • Model to understand diffusion and mass transfer to facilitate control schemes • Implement online control • Control N2O

TOPIC 3 Product Development

	Plan	Prepare & Implement	Evaluate & Improve
Marketing	*Data Collection* • Collect data to understand the market ○ Identify stakeholders and customers in the market for nutrients and other recovered resources ○ Assess competition for product ○ Track economic projections on future supply of phosphorus to keep up with current thinking about potential shortages of mined phosphorus ○ Use market-driven standards (ex. U.S. Department of Agriculture partnership for phosphorus products) ○ Evaluate potential for greenhouse gas reduction credits	*Develop Marketing Strategy* • Create a value-cost proposition ○ Identify value of recovered resource ○ Evaluate recovery costs • Develop marketing, sales, and branding strategy • Communicate benefits and advantages of nutrient recovery	*Sell Recovered Resources* • Brand and sell nutrients and other recovered resources • Sell greenhouse gas offset credits
Product Production and Control	*Product Development* • Develop a quality assurance program and process for products	*Product Production* • Produce product for market	
Public Communications	*Identify Stakeholder Values* • Identify values of community and utility board of trustees • Shift cultural mindset from "meeting the permit" to recovering resources • Identify environmental incentives for nutrient recovery	*Public Outreach* • Develop public understanding of the new purpose of a WRRF • Conduct public relations for nutrient recovery	*Share Experience* • Share best practices with other utilities and the sector

TOPIC 4 Evaluation of Alternatives

	Plan	Prepare & Implement	Evaluate & Improve
General Evaluation	*Identify Treatment Levels* • Determine level of treatment available • Determine level of treatment required or desired • Define operational/process changes required to recover nutrients	*Identify Opportunities* • Identify additional opportunities requiring more time or capital to implement, and develop a plan to finance/implement • Assess liquid vs. solid recovery (water reuse vs. land application/struvite recovery) • Ensure biosolids nutrients are considered a resource	*Evaluate Nutrient Recovery* • Identify research and development needs to drive innovations • Evaluate how simultaneous nitrification and denitrification can lower carbon requirements • Evaluate whole facility nutrient recovery • Identify water quality trading and greenhouse gas offset credit opportunities
Nitrogen Removal Evaluation	*Identify Low-Cost Options* • Determine whether the facility currently has capacity (tankage, blowers, etc.) for nitrification and denitrification • Evaluate the energy cost of various technologies and sustainable nitrogen removal for various effluent concentrations • Evaluate opportunities for high-nitrogen sidestreams	*Optimize Facility* • Reuse existing tankage to get immediate nitrogen removal to ~10-15 mg/L at low cost • Consider N2O emissions from nitrification and denitrification when designing the system	*Evaluate Long-Term Improvements* • Study impact of low dissolved oxygen and ammonia-based aeration control on: ○ PAO uptake of phosphorus ○ Competition between PAOs and glycogen-accumulating organisms (GAOs) ○ Increases in N2O emissions
Phosphorus Removal Evaluation	*Evaluate Short-Term Improvements* • Modifying existing facilities to get low-hanging fruit ○ Phosphorus to ~1 mg/L at no/low cost ○ Switch off mixers for low-tech phosphorus removal ○ More carbon for fermentation of mixed liquor • Recognize emerging information to minimize cost of chem-P ○ Good mixing at the point of application ○ Ferric/Magnesium addition	*Optimize Facility* • Recognize phosphorus-removal techniques vary for meeting different objectives, such as developing a product vs. minimizing negative impacts on receiving waterbodies	*Observe Regulatory Landscape* • Assess phosphorus limitations on land application of biosolids in areas with phosphorus saturated soils • Based on Natural Resources Conservation Service code 590, land application may not be applicable for phosphorus recovery/reuse, depending on local conditions
Carbon	*Identify Low-Cost Options* • Determine if facility can denitrify simply and at low cost (excess capacity, sufficient carbon), and what effluent concentrations/mass could be achieved	*Investigate Potential Problems* • Explore supplemental carbon issues ○ Procurement policy and pricing ○ Demonstration and testing ○ Waste products	

TOPIC 5 Risk Management

	Plan	Prepare & Implement	Evaluate & Improve
Regulatory Risk Management	*Identify and Prioritize Risks* • Evaluate legal/regulatory implications of voluntary action • Identify early technology adoption risks ○ State-of-the-art technology vs. new technology (anammox, demon, etc.)	*Mitigate Risks* • Develop strategy for risk mitigation and/or sharing ○ "Creating the Space for Innovation" ○ "Safe Harbor"	*Leverage Innovation* • Organization successfully tries to implement innovative projects and is adaptable to emerging opportunities
Upstream Management	*Understand Upstream* • Reduce inflow and infiltration • Evaluate water efficiency and conservation • Evaluate water efficiency and conservation • Separate combined systems • Pretreat to minimize nitrification inhibition • Identify sewer scalping and stream separation opportunities • Integrate resource management – include outside wastes and broaden markets	*Create Balanced Framework* • Pretreatment policy ○ Charge for Total Kjeldahl Nitrogen (TKN), total phosphorus, and low alkalinity ○ Reward for readily biodegradable chemical oxygen demand and high alkalinity • Put a water efficiency program in place • Decentralize infrastructure to optimize recovery ○ Treatment (sewer scalping) • Stream separation	*Balance Source Control with* Resource Recovery • Seek partnerships for animal waste and other waste streams • Implement sewer scalping and urine diversion • Recover proteins, and remove nitrogen and phosphorus before treatment (industrial waste) • Separate ecological sanitation streams
Innovation	*Reward Innovation* • Reduce risk through collaborative research and information sharing • Have WRRF leadership/managers recognize and reward innovative approaches • Characterize discharge impacts by conducting research on how receiving waters react to different effluent concentrations	*Improve Models* • Develop better modeling tools ○ Multiple populations of AOB and NOB ○ Kinetics for anammox bacteria (AMX) ○ Bioaugmentation speed ○ Biofilms and granules • Develop better stoichiometry and kinetics for chem-P models • Improve simulation models for GAOs and the benefit to bio-P from internal carbon fermentation • Model the digester for sidestream returns	*Pilot Test* • Test model recommendations • Obtain patents to protect the WRRF and water sector
Infrastructure Risk Management	Plan for the Future • Identify unit operations/basins for use in future iterations of nutrient reduction • Leave space in the facility hydraulic profile to accommodate future processes	Mitigate Risks • Evaluate future reduction targets • Design for current requirements with an eye toward future requirements	Evaluate Options • Assess resource recovery versus treatment/reduction

Index

CPSIA information can be obtained
at www.ICGtesting.com
Printed in the USA
LVHW060818271020
669889LV00001B/3

9 781572 783140